T0223942

Plant–Soil Slope Interaction

Plant–Soil Slope Interaction

Charles W. W. Ng

Anthony K. Leung

Junjun Ni

CRC Press
Taylor & Francis Group
Boca Raton London New York

CRC Press is an imprint of the
Taylor & Francis Group, an **informa** business

CRC Press
Taylor & Francis Group
6000 Broken Sound Parkway NW, Suite 300
Boca Raton, FL 33487-2742

First issued in paperback 2021

© 2019 by Taylor & Francis Group, LLC
CRC Press is an imprint of Taylor & Francis Group, an Informa business

No claim to original U.S. Government works

ISBN 13: 978-1-03-209129-7 (pbk)
ISBN 13: 978-1-138-19755-8 (hbk)

Library of Congress Cataloging-in-Publication Data

Names: Ng, Charles, author. | Leung, Anthony, 1985- author. | Ni, Junjun, author.
Title: Plant-soil slope interaction / Charles Ng, Anthony Leung and Junjun Ni.
Description: Boca Raton : Taylor & Francis, a CRC title, part of the Taylor & Francis imprint, a member of the Taylor & Francis Group, the academic division of T&F Informa, plc, [2019] | Includes bibliographical references and index. |
Identifiers: LCCN 2018054798 (print) | LCCN 2018056650 (ebook) | ISBN 9781351052375 (Adobe PDF) | ISBN 9781351052351 (Mobipocket) | ISBN 9781351052368 (ePub) | ISBN 9781138197558 (hardback) | ISBN 9781351052382 (ebook)
Subjects: LCSH: Soil stabilization. | Slopes (Soil mechanics) | Plant-soil relationships. | Soil-binding plants.
Classification: LCC TA749 (ebook) | LCC TA749 .N44 2019 (print) | DDC 624.1/51363--dc23
LC record available at https://lccn.loc.gov/2018054798

**Visit the Taylor & Francis Web site at
http://www.taylorandfrancis.com**

**and the CRC Press Web site at
http://www.crcpress.com**

Contents

Preface

This book deals with two natural materials – unsaturated soils and plants. Unsaturated soils cover most of the earth's land surface and support civil engineering structures. Plants are 'natural engineers' that have been stabilising initially unsaturated soil slopes and embankments for 4,000 years in both ancient China and Europe. Mr. Henry Hobhouse, who was a broadcaster, journalist, farmer, author and politician, once said '*Nature can halt our progress and nature can advance it, and man would be foolish to overlook his role......*'

Plant–Soil Slope Interaction is an interdisciplinary book that bridges the fields of unsaturated soil mechanics, plant science and ecological engineering. For those interested in applying plant-based bioengineering methods to slope stabilisation, this book provides an easy access to a collection of related findings. The contents are either extracted from the authors' own publications over the last 10 years or made verbatim copy from the publications of others, with consent of and approval from all concerned parties. Therefore, this book does not describe original findings but rather serves as a handy reference. To gain a thorough understanding of plant–soil slope interaction, readers are encouraged to first peruse the reference book *Advanced Unsaturated Soil Mechanics and Engineering*, co-authored by Charles W. W. Ng and Bruce Menzies and published by Taylor & Francis Group in 2007.

To mitigate the adverse impact of climate change and leave a better world for future generations, engineers are now looking for more environmentally friendly or 'green' solutions for slope stabilisation. Public concern over the environment and sustainability has motivated governments and engineers to gradually adopt vegetation as a kind of construction material. Vegetation can potentially offer a more environmentally friendly, cost-effective and aesthetically pleasant solution for shallow-slope stabilisation. However, the current practice considers vegetation mainly for aesthetic purposes and erosion control. The engineering functions of plant roots have been generally overlooked in the scientific analysis and design of slope stability in most parts of the world.

Historically, engineers have often been interested in the mechanical effects of roots and their biomechanical properties such as tensile strength to improve slope stability. Roots can increase the shear strength of a soil matrix through their tensile strength. On the other hand, this book provides the evidence and importance for the hydrological contributions of plant transpiration to induce soil suction. Transpiration-induced suction not only increases the shear strength of the soil but also, more importantly, decreases the water permeability, which is often ignored in the design analysis of soil slopes and embankments.

The book discusses ecological engineering and the effects of plants on the engineering performance of vital civil infrastructure systems such as slopes and embankments subjected to climate change and instability problems. The latest advanced knowledge of vegetation effects on the hydro-mechanical responses of unsaturated soils and also of the behaviour of vegetated soil slopes is presented. This is the first combined ecological-geotechnical engineering book to cover in detail not only the mechanical effects of root reinforcement but also the hydrological effects of plant transpiration on soil suction and water permeability – an area that has been less emphasised in the field. Special attention is paid to the effects of root architecture and plant spacing on induced soil suction and slope stability. The book details a fundamental understanding of soil–plant–water interaction, based on which the performance of vegetated soil slopes, including their rainfall-induced seepage and failure mechanisms, is discussed. Novel artificial root systems are created for capturing both the hydrological and mechanical effects of plant roots of different architectures in a geotechnical centrifuge. This has allowed a detailed study of the vegetation effects on slope hydrology, stability and failure mechanisms. The book also gives analytical equations for predicting the hydrological effects of plant roots on slope hydrology and stability.

This book is divided into six chapters. Chapter 1 introduces the role of vegetation in civil engineering and some basic theories underlying unsaturated soil mechanics, which are vital to understanding the rest of the book. This chapter also provides a background about the energy, water, carbon and nutrient cycles in a soil–plant–atmosphere system, as well as the water adsorption and transportation mechanisms of vascular plants.

Chapter 2 covers the hydrological effects of plants on soil suction. Specifically, the effects of soil density, plant density, plant type and carbon dioxide concentration on plant growth and transpiration-induced suction are discussed. The chapter also correlates induced soil suction with plant traits and investigates root-induced changes in soil hydraulic properties, including soil water retention curves and water permeability functions.

In Chapter 3, the mechanical effects of plant root reinforcement are discussed by revisiting the empirical power decay law and applying this law to a wider range of root diameters. The root tensile behaviour of some plant species native to Asia and the effects of plant–fungal interaction on root biomechanics are described.

Chapter 4 presents three field studies on how transpiration affects groundwater flow, slope hydrology and slope deformation. The experiments were performed on a compacted sandy ground at the HKUST Eco-Park at Tseung Kwan O in Hong Kong, a cut slope in expansive clay in Zaoyang, Hubei, and a natural saprolitic hillslope Lanta in Hong Kong.

In Chapter 5, theoretical analyses of the hydrological effects of plants with different root architectures on matric suction and slope stability are described. These analyses focus on plant transpiration-induced changes in matric suction and slope stability.

Chapter 6 discusses plant effects on slope hydrology, stability and failure mechanisms, uncovered using geotechnical centrifuge modelling techniques. First, the fundamental principles of centrifuge modelling are introduced. The development of novel artificial roots for modelling the hydro-mechanical effects of plants is then described, with particular emphasis on the root architecture. The effects of transpiration on root pull-out resistance, the effects of root architecture on the distribution of pore water pressure and plant hydro-mechanical contributions to slope hydrology and stability are revealed. The chapter ends by illustrating and discussing the effects of root architecture on slope failure mechanisms.

Charles W. W. Ng

Anthony K. Leung

Junjun Ni

Acknowledgments

We thank Dr. Ankit Garg (associate professor at Shantou University, China), Mr. Sunil Poudyal and Miss Yu-Chen Wang (postgraduate research students at HKUST) for their time and great effort to proofread and improve the quality of the entire book. The following research grants are also acknowledged for the financial support for some of the research work presented in this book:

- Research grants HKUST9/CRF/09 from the Research Grants Council of the Government of the Hong Kong SAR
- Research grants HKUST6/CRF/12R from the Research Grants Council of the Government of the Hong Kong SAR
- Research grant 2012CB719805 from the National Basic Research Program (973 Program), administered by the Ministry of Science and Technology of the People's Republic of China
- Research grant 51778166 from the Natural Science Foundation of China

In addition, we acknowledge permissions from publishers to use verbatim extracts from their published works as follows:

John Wiley & Sons, Inc., from:

- Garg, A., Coo, J. L. and Ng, C. W. W. (2015). Field study on influence of root characteristics on suction distributions in slopes vegetated with *Cynodon dactylon* and *Schefflera heptaphylla*. *Earth Surface Processes and Landforms*, 40(12), 1631–1643.
- Leung, A. K., Garg, A., Coo, J. L., Ng, C. W. W. and Hau, B. C. H. (2015). Effects of the roots of *Cynodon dactylon* and *Schefflera heptaphylla* on water infiltration rate and soil hydraulic conductivity. *Hydrological Processes*, 29(15), 3342–3354.
- Schimel, D. S. (1995). Terrestrial ecosystems and the carbon cycle. *Global Change Biology*, 1(1), 77–91.

Springer Nature, from:

- Bischetti, G. B., Chiaradia, E. A., Simonato, T., Speziali, B., Vitali, B., Vullo, P. and Zocco, A. (2005). Root strength and root area ratio of forest species in Lombardy (northern Italy). *Plant and Soil*, 278, 11–22.
- Boldrin, D., Leung, A. K. and Bengough, A. G. (2017). Correlating hydromechanical properties of vegetated soil with plant functional traits. *Plant and Soil*. doi:10.1007/s11104-017-3211-3.

- Chen, X. W., Kang, Y., So, P. S., Ng, C. W. W. and Wong, M. H. (2018). Arbuscular mycorrhizal fungi increase the proportion of cellulose and hemicellulose in the root stele of vetiver grass. *Plant and Soil*. doi:10.1007/s11104-018-3583-z.
- De Baets, S., Poesen, J., Reubens, B., Wemans, K., De Baerdemaeker, J. and Muys, B. (2008). Root tensile strength and root distribution of typical Mediterranean plant species and their contribution to soil shear strength. *Plant and Soil*, 305, 207–226.
- Leung, A. K. and Ng, C. W. W. (2016). Field investigation of deformation characteristics and stress mobilisation in a soil slope. *Landslides*, 13(2), 229–240.
- Ng, C. W. W., Kamchoom, V. and Leung, A. K. (2016). Centrifuge modelling of the effects of root geometry on the transpiration-induced suction and stability of vegetated slopes. *Landslides*, 13(5), 1–14.
- Ng, C. W. W., Leung, A. K. and Ni, J. J. (2018). Bioengineering for slope stabilisation using plants: Hydrological and mechanical effects. In *Proceedings of China-Europe Conference on Geotechnical Engineering*, Springer, Cham, Switzerland, pp. 1287–1303.
- Ng, C. W. W. (2014). The 6th ZENG Guo-xi Lecture: The State-of-the-Art centrifuge modelling of geotechnical problems at HKUST. *Journal of Zhejiang University-Science A (Applied Physics & Engineering)*, 15(1), 1–21.
- Steudle, E. (2000). Water uptake by plant roots: An integration of views. *Plant and Soil*, 226(1), 45–56.
- Wheeler, T. D. and Stroock, A. D. (2008). The transpiration of water at negative pressures in a synthetic tree. *Nature*, 455(7210), 208–212.

National Research Council of Canada, from the *Canadian Geotechnical Journal*, as follows:

- Jotisankasa, A. and Sirirattanachat, T. (2017). Effects of grass roots on soil-water retention curve and permeability function. *Canadian Geotechnical Journal*, 54(11), 1612–1622.
- Leung, A. K. and Ng, C. W. W. (2013). Analyses of groundwater flow and plant evapotranspiration in a vegetated soil slope. *Canadian Geotechnical Journal*, 50(12), 1204–1218.
- Leung, A. K., Kamchoom, V. and Ng, C. W. W. (2017). Influences of root-induced suction and root geometry on slope stability: A centrifuge study. *Canadian Geotechnical Journal*, 54(3), 291–303.
- Lim, T. T., Rahardjo, H., Chang, M. F. and Fredlund, D. G. (1996). Effect of rainfall on matric suctions in a residual soil slope. *Canadian Geotechnical Journal*, 33(4), 618–628.
- Ng, C. W. W., Leung, A. K. and Woon, K. X. (2014). Effects of soil density on grass-induced suction distributions in compacted soil subjected to rainfall. *Canadian Geotechnical Journal*, 51(3), 311–321.
- Ng, C. W. W., Liu, H. W. and Feng, S. (2015). Analytical solutions for calculating pore water pressure in an infinite unsaturated slope with different root architectures. *Canadian Geotechnical Journal*, 52(12), 1981–1992.
- Ni, J. J., Leung, A. K. and Ng, C. W. W. (2017). Investigation of plant growth and transpiration-induced suction under mixed grass-tree conditions. *Canadian Geotechnical Journal*, 54(4), 561–573.

- Sonnenberg, R., Bransby, M. F., Hallett, P. D., Bengough, A. G., Mickovski, S. B. and Davies, M. C. R. (2010). Centrifuge modelling of soil slopes reinforced with vegetation. *Canadian Geotechnical Journal*, 47(12), 1415–1430.
- Yan, W. M. and Zhang, G. H. (2015). Soil-water characteristics of compacted sandy and cemented soils with and without vegetation. *Canadian Geotechnical Journal*, 52(9), 1–14.

Elsevier, from:

- Adhikari, A. R., Gautam, M. R., Yu, Z., Imada, S. and Acharya, K. (2013). Estimation of root cohesion for desert shrub species in the Lower Colorado riparian ecosystem and its potential for streambank stabilization. *Ecological Engineering*, 51, 33–44.
- Boldrin, D., Leung, A. K. and Bengough, A. G. (2017). Root biomechanical properties during establishment of woody perennials. *Ecological Engineering*, 109, 196–206.
- Clothier, B. E. and Green, S. R. (1994). Rootzone processes and the efficient use of irrigation water. *Agricultural Water Manage*, 25, 1–2.
- Leung, A. K., Garg, A. and Ng, C. W. W. (2015). Effects of plant roots on soil-water retention and induced suction in vegetated soil. *Engineering Geology*, 193, 183–197.
- Leung, F. T. Y., Yan, W. M., Hau, B. C. H. and Tham, L. G. (2015). Root systems of native shrubs and trees in Hong Kong and their effects on enhancing slope stability. *Catena*, 125, 102–110.
- Liu, H.W., Feng, S. and Ng, C. W. W. (2016). Analytical analysis of hydraulic effect of vegetation on shallow slope stability with different root architectures. *Computers and Geotechnics*, 80, 115–120.
- Ng, C. W. W., Garg, A., Leung, A. K. and Hau, B. C. H. (2016). Relationships between leaf and root area indices and soil suction induced during drying-wetting cycles. *Ecological Engineering*, 91, 113–118.
- Ng, C. W. W., Tasnim, R., and Coo, J. L. (2018). Effects of atmospheric CO_2 concentration on soil-water retention and induced suction in vegetated soil. *Engineering Geology*, 242, 108–120.
- Ni, J. J., Leung, A. K., Ng, C. W. W. and Shao, W. (2018). Modelling hydro-mechanical reinforcements of plants to slope stability. *Computers and Geotechnics*, 95, 99–109.
- Prasad, R. (1988). A linear root water uptake mode. *Journal of Hydrology*, 99(3), 297–306.

Institution of Civil Engineers Publishing, from:

- Kamchoom, V., Leung, A. K. and Ng, C. W. W. (2014). Effects of root geometry and transpiration on pull-out resistance. *Géotechnique Letters*, 4(4), 330–336.
- Ng, C. W. W., Ni, J. J., Leung, A. K. and Wang, Z. J. (2016). A new and simple water retention model for root-permeated soils. *Géotechnique Letters*, 6(1), 106–111.
- Ng, C. W. W., Ni, J. J., Leung, A. K., Zhou, C. and Wang, Z. J. (2016). Effects of planting density on tree growth and induced soil suction. *Géotechnique*, 66(9), 711–724.

The Japanese Geotechnical Society, from:

- Ng, C. W. W. and Zhan, L. T. (2007). Comparative study of rainfall infiltration into a bare and a grassed unsaturated expansive soil slope. *Soils and Foundations*, 47(2), 207–217.
- Rahardjo, H., Satyanaga, A., Leong, E. C., Santoso, V. A. and Ng, Y. S. (2014). Performance of an instrumented slope covered with shrubs and deep-rooted grass. *Soils and Foundations*, 54(3), 417–425.

Oxford University Press, from:

- Ghestem, M., Sidle, R. C. and Stokes, A. (2011). The influence of plant root systems on subsurface flow: implications for slope stability. *Bioscience*, 61(11), 869–879.

The American Society of Plant Biologists, from:

- Lynch, J. (1995). Root architecture and plant productivity. *Plant Physiology*, 109(1), 7–13.
- Brodersen, C. R., McElrone, A. J., Choat, B., Matthews, M. A. and Shackel, K. A. (2010). The dynamics of embolism repair in xylem: In vivo visualizations using high-resolution computed tomography. *Plant Physiology*, 154(3), 1088–1095.

The Soil Science Society of America, from:

- Raats, P. A. C. (1974). Steady flows of water and salt in uniform soil profiles with plant roots. *Soil Science Society of America Journal*, 38(5), 717–722.
- Yuan, F. and Lu, Z. (2005). Analytical solutions for vertical flow in unsaturated, rooted soils with variable surface fluxes. *Vadose Zone Journal*, 4(4), 1210–1218.

The American Society of Agricultural and Biological Engineers, from:

- Camp, C. R., Karlen, D. L. and Lambert, J. R. (1985). Irrigation scheduling and row configurations for corn in the southeastern coastal plain. *Transactions of the ASAE*, 28(4), 1159–1165.

American Society for Testing and Materials International, from:

- Ng, C. W. W., Leung, A. K., Kamchoom, V. and Garg, A. (2014). A novel root system for simulating transpiration-induced soil suction in centrifuge. *Geotechnical Testing Journal*, 37(5), 1–15.

Chinese Journal of Geotechnical Engineering, as follows:

- Ng, C. W. W. (2017). Atmosphere-plant-soil interaction: Theories and mechanisms. *Chinese Journal of Geotechnical Engineering*, 39(1), 1–47.

Geotechnical Engineering Office, Hong Kong, from:

- GEO report No. 287. Detailed study of the 30 June 2007 landside on wall No. 11SE-C/R7 and the adjoining hillside at Repulse Bay Road. https://www.cedd.gov.hk/eng/publications/geo_reports/geo_rpt287.html.
- Special project report (SPR1/ 2017). Factual report on Hong Kong rainfall and landslides in 2016. https://www.cedd.gov.hk/eng/publications/geo_reports/geo_rpt330.html.

Thesis:

- Garg, A. (2015). Effects of vegetation types and characteristics on induced soil suction. PhD thesis, The Hong Kong University of Science and Technology, Hong Kong.
- Leung, T. Y. (2014). The use of native woody plants in slope upgrading in Hong Kong. PhD Thesis, The University of Hong Kong, Hong Kong.
- Woon, K. X. (2013). Field and laboratory investigations of Bermuda grass induced suction and distribution. MPhil Thesis, The Hong Kong University of Science and Technology, Hong Kong.

Authors

Charles W. W. Ng is an Associate Vice President for research and development and the CLP Holdings chair professor of sustainability in the Department of Civil and Environmental Engineering at the Hong Kong University of Science and Technology (HKUST). He obtained his PhD degree from the University of Bristol in 1993 and carried out a period of postdoctoral research between 1993 and 1995 at the University of Cambridge. He returned to Hong Kong and joined HKUST as an Assistant Professor in 1995, and rose through the ranks to become a chair professor of civil and structural engineering in 2011. He was conferred the CLP-named chair in 2017. Professor Ng was elected as the President of the International Society for Soil Mechanics and Geotechnical Engineering (ISSMGE) in September 2017 for a 4-year term.

Professor Ng was also elected as an Overseas Fellow at Churchill College, the University of Cambridge, in 2005, and a Changjiang Scholar (chair professor of geotechnical engineering) by the Ministry of Education of China, in 2010. He is a Fellow of the Institution of Civil Engineers (FICE), a Fellow of the American Society of Civil Engineers (FASCE), a Fellow of the Hong Kong Institution of Engineers (FHKIE) and a Fellow of the Hong Kong Academy of Engineering Sciences (FHKEng). Currently, he is an associate editor of the *Canadian Geotechnical Journal*, and *Landslides*, and an editorial board member of many other international journals.

Professor Ng has published over 300 SCI journal articles and 230 conference papers and delivered more than 50 keynotes and state-of-the-art reports worldwide. He also delivered the 2017 Huangwenxi Lecture (黄文熙讲座) on 'Atmosphere–plant–soil Interactions: Theories and Mechanisms' organised by the *Chinese Journal of Geotechnical Engineering* and held at Tsinghua University, Beijing. The Huangwenxi Lecture is the most prestigious named geotechnical lecture in China. Professor Ng is also the main author of two reference books: (i) *Soil-structure Engineering of Deep Foundations, Excavations and Tunnels* and (ii) *Unsaturated Soil Mechanics and Engineering*. He has supervised 50 PhD and 46 MPhil students to graduation since 1995.

Anthony K. Leung is an Assistant Professor of geotechnical engineering in the Department of Civil and Environmental Engineering at HKUST. Prior to his appointment at HKUST, he was a lecturer (2012–2016), and subsequently, he was promoted to a senior lecturer (2016–2018) in civil engineering at the University of Dundee, UK. Professor Leung is currently the associate director of the Geotechnical Centrifuge Facility at HKUST and the editor-in-chief of the Bulletin of the ISSMGE. He is an editorial board member of various major international journals, including the *Canadian Geotechnical Journal*, *Landslides* and *Proceedings of the Institution of Civil Engineers—Geotechnical Engineering*. Over the last 10 years, he has published more than 40 SCI journal articles in the field of soil bioengineering and its application to soil slope stabilisation.

Junjun Ni is a visiting Assistant Professor at HKUST. He completed his bachelor's and master's degrees at Hohai University, Nanjing, China, prior to obtaining his PhD degree at HKUST in 2017. His research interests include field monitoring, laboratory testing and numerical modelling of atmosphere–plant–soil interaction and its effects on the engineering performance of soil slopes and landfill covers. Dr. Ni has published in prominent journals such as *Géotechnique*, the *Canadian Geotechnical Journal* and *Computers and Geotechnics*.

List of notations

a	Fitting parameter in van Genuchten (1980) equation
b	$\ln(10^6)$ from Fredlund et al. (1994)
A	Effective soil cross-sectional area in Eq. (3.4)
A_c	Cross-sectional area
A_r	Root cross-sectional area at the rupture location
b_1	Fitting parameter in Eq. (5.34)
b_2	Fitting parameter in Eq. (5.34)
c'	Effective cohesion
c_r	Root cohesion
D_{10}	Grain diameter at 10% passing
D_{30}	Grain diameter at 30% passing
D_{50}	Grain diameter at 50% passing
D_{60}	Grain diameter at 60% passing
d_i	Diameter of the i-th root for Eq. (3.4)
d_r	Root diameter at the rupture location
e	Void ratio
E	Young's modulus
e_0	Void ratio of bare soil
e_a	Actual vapour pressure
E_r	Elastic modulus of artificial roots
e_s	Saturated vapour pressure
F_{air}	Inflow of sensible heat flux
F_{max}	Maximum tensile force
F_p	Pull-out resistance
F_{soil}	Soil heat flux
g	Force of gravity
G	Soil heat flux density
$g(z)$	Ability of plants to take up water at a certain depth z
g_L	Leaf conductance to water vapour
G_s	Specific gravity
h	Pressure head
H	Vertical distance to the water table
H_l	Hydraulic head in plant leaves
H_r	Hydraulic head in plant roots
h_0	Pressure head at the bottom of the slope
H_0	Vertical thickness of the slope

h_1	Positive empirical fitting coefficients in Eqs (3.5) and (3.6)
h_2	Positive empirical fitting coefficients in Eqs (3.5) and (3.6)
h_c	Capillary height
h_l	Pressure head in plant leaves
h_r	Pressure head in plant roots
i	Hydraulic gradient
I_n	Second moment of inertia
I_{plant}	Rainfall intercepted by plant
K	Coefficient of lateral earth pressure
k	Water permeability of soil
k^*	Relative water permeability k/k_s
k^*	Equals L(k^*)
K_0	Coefficient of lateral earth pressure at rest
k_0^*	Steady-state relative water permeability
k_1	Positive empirical fitting coefficients in Eqs (3.1) and (3.2)
k_2	Positive empirical fitting coefficients in Eqs (3.1) and (3.2)
K_c	Crop factor
K_p	Coefficient of passive lateral earth pressure
k_s	Saturated water permeability
k_{xylem}	Water permeability of xylem
L	Root length
L^*	Equals $L'\cos\varphi$
L'	Total perpendicular depth of slope, which is equal to the sum of L_1' and L_2'
L_0	Horizontal length of the slope in numerical simulation
L_1	Root length used in Eq. (1.11)
L_1^*	Equals $L_1'\cos\varphi$
L_1'	Perpendicular depth below the root zone
L_2	Length of plant stem in Eq. (1.11)
L_2^*	Equals $L_2'\cos\varphi$
L_2'	Perpendicular depth of the root
L_{plant}	Latent heat of evapotranspiration
L_{soil}	Latent heat of soil evaporation
m	Fitting parameter in van Genuchten (1980) equation
m_1	Fitting parameter in Gallipoli's equation (Gallipoli et al., 2003)
m_2	Fitting parameter in Gallipoli's equation (Gallipoli et al., 2003)
m_3	Fitting parameter in Gallipoli's equation (Gallipoli et al., 2003)
m_4	Fitting parameter in Gallipoli's equation (Gallipoli et al., 2003)
n	Fitting parameter in van Genuchten (1980) equation
n_r	Number of roots
N	Scale factor (model/prototype)
P	Precipitation
p'	Mean effective stress
q_0	Surface flux at steady state
q_0^*	Equals q_0/k_s
q_1^*	Equals L(q_1^*)
q_1	Surface flux at transient state
q_1^*	Equals q_1/k_s

R	Universal gas constant
R_{atmo}	Incoming atmospheric heat radiation
r_c	Radius of the centrifuge
R_{in}	Infiltrated water
R_{lr}	Heat outflux from plants and the soil surface
R_n	Net radiation intercepted by plant leaves
R_{off}	Surface runoff
R_{out}	Internal drainage
$R_{outflow}$	Outflow of heat flux
R_{plant}	Root-water uptake
R_s	Capillary radius
R_{solar}	Incoming solar radiation
R_v	Root volume ratio
S	Sink term representing root water uptake
s'	Laplace-transform complex variable
S_r	Degree of saturation
t	Elapsed time
T	Transpiration rate
T	Time for dynamic condition
T_{diff}	Time for diffusion
T_a	Absolute temperature
T_{air}	Air temperature
T_r	Root tensile strength
T_s	Surface tension
T_{soil}	Soil evaporation
u	Wind speed
u_a	Pore-air pressure
u_v	Partial pressure of pore-water vapour
$\overline{u_a}$	Average pore-air pressure
$\overline{u_w}$	Average pore-water pressure
u_{v0}	Saturation pressure of pore-water vapour
u_{v1}	Partial vapour pressure above the soil water
u_w	Pore-water pressure
v	Water flow velocity
v_{w0}	Specific volume of water
w_c	Angular velocity of the centrifuge
w_{opt}	Optimum moisture content
x^*	Integration variable
x'	Coordinate parallel to the slope surface
y	Dummy variable for integration of Ψ in Eq. (1.3)
Z	Vertical coordinate with upwards positive
z	Vertical depth
z^*	Equals $z'\cos\varphi$
z'	Coordinate perpendicular to the slope surface
α	Desaturation coefficient of soil
β	Contact angle indicating the degree of soil hydrophilicity
γ	Psychometric constant

γ_t	Bulk unit weight
γ_d	Dry unit weight of soil
γ_w	Unit weight of water
δ'	Effective interface friction angle
η_b	Coefficient of horizontal subgrade reaction
θ	Volumetric water content
θ'	First derivative of θ for Eq. (1.3)
θ_r	Residual volumetric water content
θ_s	Saturated volumetric water content
λ_n	nth positive root of the equation $sin(\lambda L^*) + (2\lambda/a)\cos(\lambda L^*) = 0$
μ	Poisson's ratio
ξ	Material parameter in Eq. (6.2)
ρ_w	Water density
ρ_{max}	Maximum dry density
σ_D	Total horizontal stress
σ_D'	Effective horizontal stress
σ_n	Total normal stress
σ'_t	Tensile strength of artificial roots
$\overline{\sigma_b}$	Average effective horizontal stress
$\overline{\sigma_v}$	Average vertical stress
$\overline{\sigma'_v}$	Average effective vertical stress
τ_f	Shear strength of soil
φ	Slope angle
χ	Bishop's parameter
ϕ'	Effective friction angle
ϕ_d	Dilation angle
ϕ^b	Angle indicating the rate of increase in shear strength relative to negative pore water pressure
ϕ_{cr}'	Critical-state friction angle
ψ	Matric suction
ψ_d	Matric suction beyond which plant roots experience increasing difficulties in extracting moisture from the soil
ψ_w	Matric suction corresponding to the wilting point beyond which plant roots could no longer extract moisture from the soil
ψ_T	Total suction
ω_v	Molecular mass of water
Γ_{air}	Heat outflux from the atmosphere
Γ_{plant}	Heat outflux from plant
Γ_{soil}	Heat outflux from soil
Δ	Slope of the vapour pressure curve
ΔS_{air}	Air heat variations due to air temperature change
ΔS_{plant}	Heat variation of plant caused by changes in the plant and soil temperatures
ΔS_{reac}	Plant heat variations due to plant temperature change
ΔW_{plant}	Water stored in plant
ΔW_{pond}	Water ponding
ΔW_{soil}	Water stored in soil

List of nomenclature

AET	Actual evapotranspiration
AEV	Air entry value
AMF	Arbuscular mycorrhizal fungi
ANCOVA	Analysis of covariance
AT	Actual transpiration
CA	Cellulose acetate
CO_2	Carbon dioxide
CDG	Completely decomposed granite
CDT	Completely decomposed coarse ash tuff
CesA	Cellulose synthase
CP	Piezometers
CSL	Critical-state line
DMT	Dilatometer test
ET	Evapotranspiration
Fm	*Funneliformis mosseae*
FOS	Factor of safety
Ga	*Glomus aggregatum*
GAX	Glucuronoarabinoxylan
GMR	Green mass ratio
GWT	Groundwater table
HDS	Heat dissipation sensor
HKUST	Hong Kong University of Science and Technology
HRGP	Hydroxyproline-rich glycoprotein
JFT	Jet-Filled tensiometer
K	Potassium
LAI	Leaf area index
LL	Liquid limit
Me	*Melastoma sanguineum*
ML	*Silt*
MS	Matric suction
N	Nitrogen
NM	Non-mycorrhizal

P	Phosphorous
PC	Principal-component
PE	Potential evaporation
PET	Potential evapotranspiration
PI	Plasticity index
PL	Plastic limit
PH	Plant height
PR	Penetration resistance
PT	Potential transpiration
PWP	Pore-water pressure
RAI	Root area index
RAR	Root area ratio
RB	Root biomass
RC	Relative compaction
Re	*Reevesia thyrsoidea*
Rh	*Rhodomyrtus tomentosa*
RH	Relative humidity
Ri	*Rhizophagus intraradices*
RLD	Root length density
RSR	Root-to-shoot ratio
SB	Shoot biomass
Sc	*Schefflera heptaphylla*
SC	Clayey sand
SLA	Specific leaf area
SP	Standpipe
SPSS	Statistical Package for the Soil Sciences
SRL	Specific root length
SRM	Strength reduction method
SWRC	Soil water retention curve
TE	Transpiration efficiency
Tot B	Total biomass
VWC	Volumetric water content
USCS	Unified Soil Classification System
WD	Wood density

Chapter I

Introduction

1.1 ROLE OF VEGETATION IN CIVIL ENGINEERING

For many centuries, living plants and dead wood had been used to reinforce soil slopes, embankments, foundations (e.g., timber piles) and earthen retaining walls such as those in ancient China (Smith and Snow, 2008) and ancient Rome (Partov et al., 2016). The design of these green reinforcement technologies was essentially empirical. As the world became increasingly industrialised, concrete and steel replaced timbers as the key materials in various types of infrastructural development and construction projects, including the improvement of slope stability. The mechanical properties of these man-made materials are highly controllable and predictable, and hence, they offer engineers and designers a better sense of safety and security in general. Nowadays, however, people seek for more environmentally friendly and green solutions to many engineering problems. The desire of the public to create a sustainable world for future generations has gradually motivated governments and engineers to rediscover vegetation as an engineering material. Soil bioengineering using plants can potentially offer an environmentally friendly, cost-effective and aesthetically pleasant solution for enhancing the stability of shallow soil slopes and controlling the surface erosion resulting from blowing winds and moving waters.

In the tropical and subtropical regions of the world, shallow slopes (i.e., those less than 2 m deep) often fail because of the short-duration but intensive rainfalls (GEO, 2011; Ng et al., 2016a). Stone walls and chunam covers (Figure 1.1) are commonly used to protect slopes in some parts of the world, including Hong Kong. However, it has become evident that stone walls and chunam covers could not maintain the slope stability, and these methods are in fact not environmentally friendly, not to mention aesthetically unpleasant. Although vegetation has been used empirically for decades in slope protection, it is mainly done for aesthetic purposes (Coppin and Richards, 1990). In fact, the function of vegetation has not yet been fully integrated into the analysis and design of slopes in a scientific manner.

A well-recognised effect of roots on slope stability is their mechanical reinforcements in shallow soil slopes (Wu et al., 1979; Pollen and Simon, 2005). Plant roots, which can sustain tension, occupy the space of soil pore and increase the tensile strength and shear strength of soil–root composites. In the past decades, mechanical root reinforcement has been extensively quantified, both experimentally and analytically (Wu et al., 1979; Pollen and Simon, 2005; Fan and Su, 2008; Jotisankasa and Taworn, 2016). This mechanical effect can be readily included in slope stabilisation calculations (Greenwood et al., 2004; Genet et al., 2010; Mao et al., 2014). How much mechanical reinforcement contributes to soil shear strength depends strongly not only on the biomechanical properties of roots (i.e., root tensile strength and Young's modulus) but also on the root architecture and the amount of roots in the rooted zone (Pollen and Simon, 2005; Ng et al., 2016a; Boldrin et al., 2017a).

(a)   (b)

Figure 1.1 Traditional slope failure in Hong Kong: (a) a stone wall (Repulse Bay Road; 30 June 2007) and (b) a chunam cover (South Lantau Road; 7 September 2016). The two photos are provided by the Geotechnical Engineering Office, Civil Engineering and Development Department, Hong Kong SAR.

In addition to providing mechanical reinforcement, living roots induce soil matric suction (or negative pore water pressure) through root water uptake via transpiration (Ng et al., 2013; Garg et al., 2015b; Leung et al., 2015a). Soil suction can also be induced or further increased by evaporation from the soil surface. The combined physical processes of transpiration and evaporation are referred to as evapotranspiration (*ET*) (Gardner, 1960). This further enhances the shear strength of soil and more importantly helps to reduce water permeability and infiltration (Ng and Menzies, 2007; Leung and Ng, 2013a). The key to applying this green engineering technique successfully is to first understand the fundamentals of unsaturated soil mechanics and soil–plant–atmosphere interactions, which are interdisciplinary subjects involving atmospheric science, soil science, botany and geotechnical engineering.

1.2 FUNDAMENTALS OF UNSATURATED SOIL MECHANICS

To improve our understanding of soil–plant–atmosphere interactions, the relevant theories of unsaturated soil mechanics are introduced in this section. Soil suction is defined as the free energy state of soil water (Edlefsen and Anderson, 1943; Richards, 1965). According to the thermodynamic theory, the total suction ψ_T in soil is related to the partial pressure of pore–water vapour (Aitchison, 1965) and expressed as follows:

$$\psi_T = -\frac{RT_a}{\upsilon_{wo}\omega_v}\ln\left(\frac{u_v}{u_{vo}}\right) = -\frac{RT_a}{\upsilon_{wo}\omega_v}\left[\ln\left(\frac{u_v}{u_{v1}}\right) + \ln\left(\frac{u_{v1}}{u_{vo}}\right)\right] \tag{1.1}$$

where R is the universal gas constant, T_a is the absolute temperature, ω_v is the molecular mass of water, υ_{wo} is the specific volume of water, u_v is the partial pressure of pore–water vapour, u_{v1} is the partial vapour pressure above the soil water and u_{vo} is the saturation pressure of pore–water vapour over a flat surface of pure water at the same temperature (Fredlund and Rahardjo, 1993). According to Eq. (1.1), the total soil suction is composed of soil matric suction (the first term on the right-hand side of the equation) and osmotic suction (the second term on the right-hand side of the equation).

Up to now, soil–plant–atmosphere interactions have not been taken into account in the design of geotechnical infrastructure, mainly due to limited understanding of the complex mechanisms involved. Plants would reduce the soil water content, and the corresponding increase in suction would lead to changes in both the shear strength and water permeability of the unsaturated soil. The effects of root water uptake on soil shear strength and water permeability are referred as 'hydrological effects' (Ng, 2017).

For simplicity, the shear strength of an unsaturated soil, τ_f, may be expressed in terms of water content and soil matric suction, as follows (Vanapalli et al., 1996):

$$\tau_f = c' + (\sigma_n - u_a)\tan\phi' + (u_a - u_w)\left[(\tan\phi')\left(\frac{\theta - \theta_r}{\theta_s - \theta_r}\right)\right]$$

(1.2)

where c' is the effective cohesion; σ_n is the normal stress; u_a and u_w are the pore–air pressure and pore–water pressure, respectively; ϕ' is the effective friction angle; θ is the volumetric water content; θ_s is the saturated volumetric water content and θ_r is the residual water content. The difference between u_a and u_w (i.e., $u_a - u_w$) is called matric suction. More advanced theories such as effects of state-dependent dilatancy on the shear strength of an unsaturated soil are given by Chiu and Ng (2003), Ng and Menzies (2007) and many others. According to Eq. (1.2), plant-induced matric suction would increase soil shear strength owing to a reduction in water content (or an increase in soil matric suction). The laboratory findings by Ng and Zhou (2005) also show that matric suction would increase the tendency of soil dilation, which in turn would improve soil shear strength.

Unlike that in the saturated soils, water permeability in an unsaturated soil not only depends on the void ratio (e) but is also governed by the degree of saturation or the water content, which in turn affects matric suction in the soil. The relationship between water permeability and matric suction can be expressed as follows (Fredlund et al., 1994):

$$k(\psi) = k_s\left[\frac{\int_{\ln(\psi)}^{b} \frac{\theta(e^y) - e(\psi)}{e^y}\theta'(e^y)dy}{\int_{\ln(\psi_{ave})}^{b} \frac{\theta(e^y) - \theta_s}{e^y}\theta'(e^y)dy}\right]$$

(1.3)

where ψ is the matric suction, $b = \ln(10^6)$, k_s is the saturated water permeability, y is a dummy variable for integration of ψ and θ' is the first derivative of function θ. Figure 1.2 shows schematic water permeability functions for different types of soil. Water permeability generally reduces with increasing matric suction. In unsaturated soil, pore air is not conductive to the movement of liquid water. Thus, an increase in suction caused by a reduction of

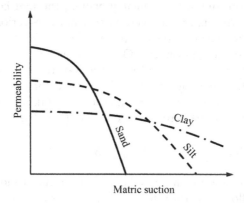

Figure 1.2 Relationship between water permeability and matric suction.

water content can lead to a decrease in water permeability. Based on Eqs (1.2) and (1.3), it is clear that plant-induced suction not only increases soil shear strength but also reduces water permeability, which in turn reduces rainfall infiltration and hence potentially helps to preserve a greater amount of soil suction in an unsaturated soil slope. The hydrological effects due to plant root water uptake had not been fully recognised until a comprehensive research was conducted in the recent few years (e.g., Ng et al., 2013, 2014a, 2015, 2016a, 2016c, 2016e; Ng, 2017).

1.3 ENERGY BALANCE, WATER, CARBON AND NUTRIENT CYCLES IN A SOIL–PLANT–ATMOSPHERE SYSTEM

Hydrological effects are attributed to plant transpiration, which is governed by moisture exchange between the atmosphere, plants and soils. Hence, plant growth is one of the most important factors determining the efficiency of moisture exchange (White and Brown, 2010). Plants require light, water and nutrients for producing glucose, proteins, lipids, nucleic acids and other carbohydrates for their life cycle and are they able to offer hydrological effects for slope stability improvement. This section describes the physiological process through which plants absorb water, carbon and nutrients. Energy flow, moisture, carbon and nutrients cycles in the soil–plant–atmosphere system are also introduced briefly.

1.3.1 Energy balance

Light is required during the processes of plant photosynthesis, water evaporation and plant transpiration. As any energy variation can cause water to change phase, such as from liquid to gas (i.e., water evaporation), the energy balance and the water balance are closely interrelated. Plant photosynthesis greatly affects plant growth (Nobel, 2009). Both soil evaporation and plant transpiration are important not only for agricultural science (such as irrigation scheduling) but also for slope reinforcement due to the induction of soil matric suction. The energy consumed by these two processes can be measured and calculated. The amount of soil evaporation and that of plant transpiration can then be estimated based on the Penman equation (Penman, 1948) and the Penman–Monteith equation (Allen et al., 1998), respectively.

Figure 1.3 shows the main components of heat transfer in an idealised soil–plant–atmosphere system (ABCD). The heat transfer is composed of heat conduction, convection and radiation. Above the ground surface, the main energy components are solar radiation, atmospheric heat radiation and sensible heat flux (i.e., thermal conduction between the ground surface and the atmosphere, as well as thermal convection caused by wind and turbulence). Because of the albedo of the ground and leaf surfaces, some of the solar radiation would be reflected to the atmosphere. On the other hand, both soils and plants can release energy into the atmosphere in the form of long-wave thermal radiation (Jones and Rotenberg, 2001). Based on the energy balance equation for the soil–atmosphere system proposed by Blight (1997), a newly modified equation that takes plants into account for the area of ABCD in the figure can be expressed as follows (Ng, 2017):

$$(R_{solar} + R_{atmo} + F_{air}) - R_{outflow} = \Delta S_{air} + \Delta S_{plant} + \Delta S_{reac} \tag{1.4}$$

where R_{solar} and R_{atmo} are the incoming solar radiation and atmospheric heat radiation, respectively; F_{air} is the inflow of sensible heat flux; $R_{outflow}$ is the outflow of heat flux; ΔS_{air} is the air heat variations due to air temperature change; ΔS_{plant} is the plant heat variations due

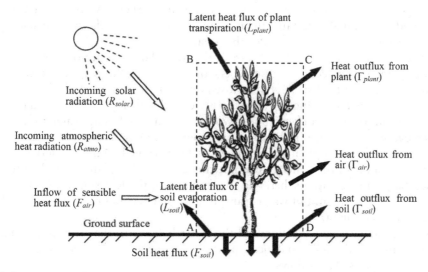

Figure 1.3 Components of the energy balance in an atmosphere-plant-soil system.

to plant temperature change and ΔS_{reac} is the heat variations induced by photosynthesis and respiration. It can be seen from Eq. (1.4) that some of the net heat inflow induces changes in the air and plant temperatures, while some is stored inside the system via plant metabolism (energy absorption during photosynthesis and heat generation during respiration). The outflow of heat $R_{outflow}$ in Eq. (1.4) can be expressed as follows:

$$R_{outflow} = \Gamma_{plant} + \Gamma_{soil} + \Gamma_{air} + F_{soil} + L_{plant} + L_{soil} + R_{lr} \tag{1.5}$$

where Γ_{soil}, Γ_{plant} and Γ_{air} are the heat outflux from soil, plants and the atmosphere, respectively; F_{soil} is the soil heat flux, which is used for the metabolism and respiration of bacteria and fungi; and L_{plant} and L_{soil} are the latent heat of transpiration and soil evaporation, respectively. Latent heat is the thermal energy released or absorbed by a body or a thermodynamic system during a constant-temperature process – usually a first-order phase transition. R_{lr} is the long-wave heat outflux from plants and the soil surface. In the equation, L_{soil} and F_{soil} affect the amount of water evaporation and hence soil suction. L_{plant} affects plant transpiration and hence suction. Different from Blight (1997), Eqs (1.4) and (1.5) consider the effects of plants on the energy balance (such as ΔS_{plant}, ΔS_{reac}, L_{plant} and Γ_{plant}).

1.3.2 Water balance

Water is an essential component of all living organisms. Water on earth is distributed in oceans, lakes, rivers, glaciers, the atmosphere and soils. The total amount of water found on earth is about 1.4×10 km³ (USGS, 2016). The ocean contains 96% of that water in the form of salty water (USGS, 2016). The remaining 4% water exists on land in the form of fresh water. More than 68% of the fresh water occurs in the form of ice in glaciers. The remaining 30% or so of the fresh water is found in the soil. Surface freshwater sources such as rivers and lakes account for about 1/15,000th of the total amount of water on earth (USGS, 2016). The water cycle on earth is continuous and dynamic. Water exchange between land and ocean forms the core of this cycle. Under solar radiation, water in the ocean and terrestrial surface evaporates, and the water vapour enters the atmosphere. A portion of that vapour is precipitated as rain under certain climate conditions. The rainfall would become surface and sub-surface water flows destined for the ocean again.

In the water cycle, the interaction between the soil, plant and atmosphere is an important research area for geotechnical researchers and engineers. Many geotechnical problems such as slope failure and debris flow are caused by rainfall. Figure 1.4 shows the main factors governing the water balance of a typical vegetated soil slope. When rainfall intensity exceeds water infiltration, surface runoff would be generated. The water infiltrating the subsurface causes an increase in the soil water content and an associated reduction in soil suction. In vegetated soils, plants consume water for various biophysical activities such as photosynthesis. Some of the water that is not consumed by the plants would be transpired or released back into the atmosphere via the stomata on the leaf surface. McElrone et al. (2013) found that the amount of water transpired by plants could be up to 95% of the total amount of root water uptake.

Considering the mass conservation and the water balance in the atmosphere–plant–soil system in Figure 1.4, the water balance equation near the ground surface (e.g., in the area EFGH) can be expressed as follows:

$$(P - I_{plant} - R_{off} - \Delta W_{pond}) + (R_{in} - R_{out}) - (T_{soil} + T_{plant}) = \Delta W_{soil} + \Delta W_{plant} \tag{1.6}$$

where P is the precipitation, I_{plant} is the rainfall intercepted by plant, R_{off} is the surface runoff, ΔW_{pond} is the water ponding, R_{in} is the infiltrated water, R_{out} is the internal drainage, T_{soil} is the soil evaporation, T_{plant} is the plant transpiration, R_{plant} is the root water uptake and ΔW_{soil} and ΔW_{plant} are the waters stored in soil and plants, respectively. The equation suggests that the net water flow into a vegetated soil is equal to the water stored in the soil and plants.

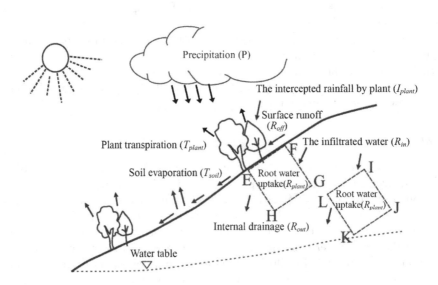

Figure 1.4 Components of the soil water balance.

For the water balance in depths (e.g., in the area IJKL in Figure 1.4, which is away from the soil surface but within the root zone), the water balance equation can be expressed as follows:

$$R_{in} - (R_{plant} + R_{out}) = \Delta W_{soil} + \Delta W_{plant} \tag{1.7}$$

The term on the left-hand side of Eq. (1.7) is the sum of net water influx and net water outflux, which includes root water uptake and internal drainage. On the right-hand side, the terms denote the net water stored in the soil and plants. Unlike Eqs (1.6) and (1.7) does not include the net rainfall influx (i.e., the term $[P - I_{plant} - R_{off} - \Delta W_{pond}]$). If the area is below the root zone, the plant effects (R_{plant} and ΔW_{plant}) can be neglected in the water balance equation. Eqs (1.6) and (1.7) describe the overall water balance in the soil–plant–atmosphere system, and they form the basis for the equations developed in the following chapters.

1.3.3 Carbon cycle

Carbon cycling in the ecosystem has direct effects on plant growth (Heimann and Reichstein, 2008). For example, the amount of carbon dioxide in the atmosphere influences plant photosynthesis and respiration. Figure 1.5 shows the total carbon reserve on earth in billion tons

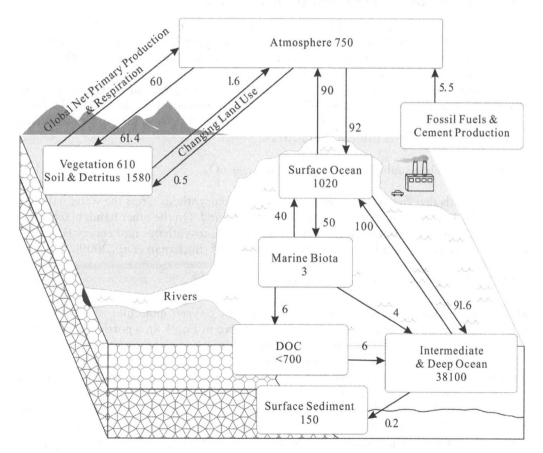

Figure 1.5 Carbon cycling on a global scale. Numbers are average values. Unit: billion tons per year. (From Schimel, D. S., *Global Change Biol.*, 1, 77–91, 1995.)

per year, as reported by Schimel (1995). The intermediate and deep oceans store most of the carbon, followed by the surface ocean, soil, detritus, the atmosphere and vegetation. Among them, vegetation, soil and detritus are the major carbon sinks on land (i.e., the terrestrial environment). Over the years, various attempts have been made to increase the carbon content lockup in the soil to reduce the concentration of carbon dioxide (a greenhouse gas) in the atmosphere (Lal, 2004; McCarl et al., 2006; Fontaine et al., 2007; Washbourne et al., 2015).

In terrestrial ecosystems, plants and microorganisms influence the amount of carbon in soil. Plants capture carbon dioxide and fix the carbon molecules through photosynthesis, while most of the microorganisms release it through respiration. Plants also release carbon dioxide through respiration. Carbon dioxide and water are consumed to produce photosynthates (i.e., glucose, a simple sugar with the molecular formula $C_6H_{12}O_6$), some of which are then stored or synthesised in the body (e.g., starch and cellulose), while the rest are consumed to maintain other plant metabolic processes. The consumed photosynthates during the metabolic processes are delivered to the soil in the form of root exudates and then utilised by microorganisms (e.g., rhizospheric bacteria), while some are acquired by symbiotic fungi and further transferred to the soil in the form of glomalin (Smith and Read, 2008). Organic debris in the soil is decomposed by microbes, and then, the carbon within is released back into the atmosphere. This completes the carbon cycle.

1.3.4 Photosynthesis and respiration of plants

Photosynthesis is the process by which plants, algae and some bacteria convert carbon dioxide, water or hydrogen sulphide into carbohydrates by using light energy (Gates, 1980). Plant chloroplasts convert and store the energy from sunlight (specifically blue light with a wavelength of 425–450 nm and yellow and red light with a wavelength of 600–700 nm) in energy-storing molecules ATP and NADPH while freeing oxygen from water (taken up by roots). Subsequently, chloroplasts use ATP and NADPH to produce organic molecules from carbon dioxide (entered the plant through leaf stomata) in a process known as the Calvin cycle, described as follows (Moran et al., 2011):

$$6H_2O + 6CO_2 + \text{sunlight} \rightarrow C_6H_{12}O_6 \text{ (glucose)} + 6O_2 \qquad (1.8)$$

Eq. (1.8) reveals that the use of water by plants for photosynthesis alters the water balance, described by Eq. (1.6), and hence soil suction in the ground. On the other hand, plant respiration is the process through which plants transform photosynthates into energy (Eq. 1.9), which is necessary for all life processes (i.e., metabolism) (Buchanan et al., 2009).

$$C_6H_{12}O_6 + 6O_2 \rightarrow 6CO_2 + 6H_2O + \text{energy} \qquad (1.9)$$

The oxygen consumed for respiration comes from the oxygen molecules in water generated during photosynthesis, as well as the air. As shown in Eq. (1.8), a portion of the water absorbed by plants provides energy for generating glucose for plant growth. Another portion, albeit a small one, would be stored in the plants (ΔW_{plant} in Eqs [1.6] and [1.7]). The remainder would be delivered from the soil to the atmosphere via plant transpiration (i.e., T_{plant} in Eq. [1.6]) (Buchanan et al., 2009).

According to Eqs (1.6) through (1.9), water and carbon are recycled with the help of plants. The water evaporated from the soil to the atmosphere returns to the plant–soil system via precipitation (Eq. 1.6). Besides photosynthesis, respiration is also essential for plant survival, and it involves the conversion of glucose into energy. Carbon dioxide and water generated by respiration are then released into the atmosphere (see Eq. [1.9]). The difference in the daytime and nighttime transpiration rates creates a difference in the amount of root water uptake between day and night.

1.3.5 Nutrient cycles

Nutrients are essential for plant growth. Nutrients in the ecosystem consist of macronutrients, including nitrogen, phosphorus, potassium, sulphur and other micronutrients (such as magnesium, zinc and iron). Poor plant growth conditions may reduce the extent to which plants are able to reinforce soil.

Nitrogen is the basic element of many biological compounds, such as protein. As much as 78% of air is composed of nitrogen gas. According to Galloway et al. (2008) and Lu et al. (2011), the nitrogen cycle is the physical cycle through which nitrogen and nitrogen compounds convert into each other in nature. Nitrogen exists in all amino acids that form proteins and is one of the four basic elements that make up nucleic acids such as DNA (Deoxyribonucleic acid). As shown in Figure 1.6, plants consume large amounts of nitrogen to produce chlorophyll for photosynthesis and growth. Processing (or fixation) is the inevitable process of converting gaseous free nitrogen into nitrogen compounds (e.g., nitrate) that are available for (bioavailable) organisms (Bloom et al., 2002). A small portion of nitrogen is fixed by lightning, but most of it is fixed by non-symbiotic or symbiotic nitrogen bacteria. These bacteria have nitrogen-fixing enzymes that promote the synthesis of nitrogen and hydrogen into ammonium, which is then transformed by the bacteria to form part of their tissues (Postgate, 1998). Some nitrogen-fixing bacteria such as rhizobium develop in the nodules of legumes (e.g., peas and faba beans) and establish a mutualistic relationship with plants, producing ammonium for the plants and obtaining sugars in return. Therefore, land can be made more fertile by planting legumes. The addition of nitrogen fertilisers (in the form of nitrate or ammonium) also affects the nitrogen cycle (Postgate, 1998).

The absorption of nitrate or ammonium by plants decreases the osmotic suction in soil. The sum of osmotic and matric suction gives the total suction (Eq. 1.1). In nature, osmotic

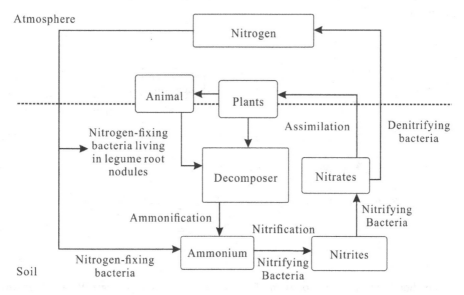

Figure 1.6 Nitrogen cycling in an ecosystem. (From Galloway, J. N. et al., *Science*, 320, 889–892, 2008.)

suction is continuously changing. With less precipitation, the salt concentration in soil would increase, resulting in a high osmotic suction.

A lack of nitrogen would cause a deficiency of chlorophyll in leaf and hence leaf yellowing. This affects plant photosynthesis and respiration and hinders plant growth (reduced biomass such as leaves, roots and fruits). Reduced plant biomass due to the lack of nitrogen would generally reduce root water uptake (Andrews et al., 2013; Ng et al., 2016e), potentially affecting the soil shear strength, water permeability and consequently slope stability (Eqs [1.2] and [1.3]).

Figure 1.7 shows a typical phosphorus cycle in an ecosystem (Stevenson and Cole, 2008). Dissolved phosphorus would be stored in the soil owing to weathering, chemical reactions, artificial additives, atmospheric deposition and degradation of the organic matter (i.e., plants and animals). The dissolved phosphate would increase the osmotic suction of the soil (Ng, 2017), potentially improving slope stability and preventing erosion more effectively. Phosphorus also becomes a part of organisms through plant absorption and microbial fixation (Jansson, 1988; Schachtman et al., 1998). In plants, phosphorus is used to synthesise DNA molecules and proteins, which are essential for the growth of new cell tissues, as well as for the transformation of energy within plants. Plants lacking phosphorus are normally short and stunted (Schachtman et al., 1998). The lack of phosphorus would reduce plant photosynthesis and negatively affect respiration.

Sulphur is recycled similarly in the ecosystem (Figure 1.8). According to Kellogg et al. (1972), sulphur exists mainly in the form of sulphate, elemental sulphur and organic sulphur in the soil, as well as in the form of sulphur dioxide and sulphur suspended matter in the air. These various forms of sulphur convert into each other through biological and chemical reactions. Sulphur in the air would enter the soil (e.g., volcanic ash), forming sulphate. Sulphate is then absorbed by plants or is converted into elemental sulphur by bacteria. Some of the sulphate would be discharged into rivers and sea through water flow. Plants absorb sulphate from the soil and transform it into organic sulphides, which are then consumed and excreted by animals, consequently creating another form of organic sulphides. Organic sulphides are again decomposed by bacteria and transformed into sulphate before entering the soil again (Kellogg et al., 1972; Jørgensen, 1977). Sulphur is a component of proteins, hormones and vitamins in plants. A lack of sulphur would cause leaf yellowing and hence affect plant growth.

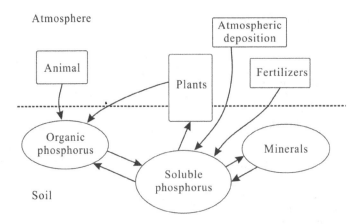

Figure 1.7 Phosphorus cycling in an ecosystem. (From Cross, A. F. and Schlesinger, W. H., *Geoderma*, 64, 197–214, 1995.)

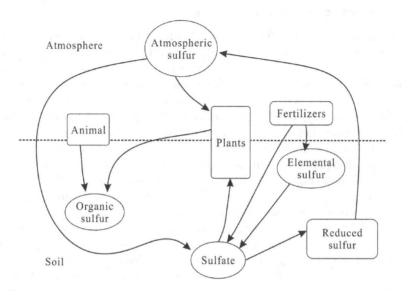

Figure 1.8 Sulfur cycling in an ecosystem. (From Lovelock, J. E. et al., *Nature*, 237, 452, 1972.)

The cycles of energy, carbon, water and nutrients are important for animals, plants and microorganisms to coexist. In turn, these organisms also affect the cycles of energy, carbon, water and nutrients. To promote plants for engineering purposes, the role of plants in the earth's ecosystem and their mechanisms must be understood comprehensively.

1.4 WATER ABSORPTION AND TRANSPORTATION MECHANISM OF VASCULAR PLANTS

There are more than 280,000 identified plant species, including trees, shrubs, climbers and ferns (http://www.theplantlist.org/). These plants can be categorised into vascular or non-vascular ones. Vascular plants, including trees, shrubs and climbers, use vascular tissues such as xylem and phloem to transport water and nutrients from the roots to other parts of the plant (such as the stem and leaves). Non-vascular plants such as algae are not equipped with such tissues. In general, only vascular plants are used in slope engineering, as they have extensive water-absorbing root systems that can offer both mechanical and hydrological reinforcement. Therefore, only their water absorption mechanisms are introduced in this book, and they are simply referred to as 'plants'.

Among all abiotic factors, water is usually the most limiting one. Most of the water obtained by plants is absorbed from the soil through roots (McElrone et al., 2013). Plants need water for photosynthesis. Carbon dioxide in the atmosphere enters plants, while at the same time, water is lost via their stomata. According to a previous study by McElrone et al. (2013), transpiring 400 water molecules would produce only one molecule of carbon dioxide. Thus, plants must absorb water continuously to obtain enough carbon dioxide for photosynthesis. Plants transpire water not only during the day but also throughout the night. However, the transpiration rate during the nighttime is typically 5%–15% of that during the daytime. The transpiration rate is related to soil water content, wind and humidity conditions (Caird et al., 2007; Dawson et al., 2007).

1.4.1 Mechanisms of root water uptake

Plant roots of various lengths and diameters together form a complex system. Roots grow from the root tip, forming non-woody fine roots, which form the most permeable portion of a root system (McCully, 1999). The fine roots are mainly responsible for absorbing water and nutrients from the soil. Numerous root hairs are distributed on the surface of fine roots. They drastically increase the surface area for water uptake and improve the contact between the soil particles and the fine roots (McCully, 1999). Plants maintain a proper water balance by continuously adjusting the water conductance in the plant cells (Maurel and Chrispeels, 2001; Maurel et al., 2008). Under dry condition, embolism (cavitation) would occur in the vessels where water flows. Embolism can be self-repaired by drawing fluid from the surrounding fibres and parenchyma cells into the vessel (Brodersen et al., 2010). Under flooded condition, plant roots would be malfunctioned in water uptake owing to the lack of oxygen in the root zone. For wetland plants (e.g., mangroves), oxygen is transported from shoot to root for root functioning. Root exudates enhance the soil ability of water retention, which resists drought stress (Carminati et al., 2016). Plants can also promote root water uptake by developing a symbiotic relationship with fungi (e.g., arbuscular mycorrhizal fungi). More details are provided in Chapter 3.

Figure 1.9 shows three possible pathways of water transport from soil into the root xylem, namely (i) the apoplastic pathway (within the cell wall continuum), (ii) the symplastic pathway (through cytoplasmic continuities and plasmodesmata) and (iii) the transcellular pathway (mediated by aquaporins located within the plasma membranes) (Steudle, 2000; Maurel et al., 2008). Long-distance transportation of water from roots to other parts of the plant (e.g., leaves) occurs mostly through the xylem in the absence of membrane barriers. Among the water transport mechanisms described in Figure 1.10, the apoplastic and symplastic pathways are driven by xylem pressure, whereas the transcellular pathway is driven by osmotic suction (Maurel et al., 2008).

1.4.2 Mechanisms of water transport from roots to leaves

Once water enters the fine roots, it migrates towards the centre of the root by passing through the epidermis, cortex and endodermis before finally reaching the xylem. Figure 1.10 shows a highly idealised water flow process inside a plant system. It is generally recognised that water is driven and transported through the xylem by the hydraulic gradient. When the

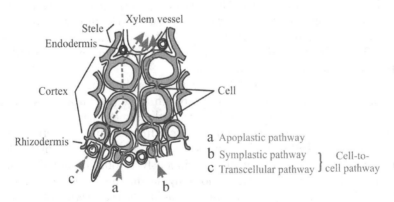

Figure 1.9 Cross-section of transportation pathways of water and solutes inside roots. (From Steudle, E., *Plant Soil*, 226, 45-56, 2000.)

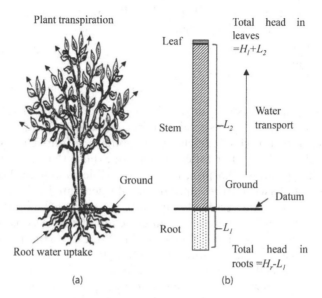

Figure 1.10 Schematic diagram of (a) water transportation from plant roots to leaves and (b) an idealised water transportation model. (Simplified from Wheeler, T. D. and Stroock, A. D., *Nature*, 455, 208–212, 2008.)

water arrives at the stem, it is delivered to leaf veins and then distributed to mesophyll cells (where photosynthesis takes place) (Sack and Tyree, 2005). The xylem, which is the water-conducting tissue of the plant, can be conceived as a porous medium (Siau, 1984). The water flow inside the xylem can then be described by Darcy law:

$$v = -k_{xylem}i \tag{1.10}$$

where v is the water flow velocity, k_{xylem} is the water permeability of xylem and i is the hydraulic gradient between plant roots and leaves, which can be expressed as follows:

$$i = \frac{(H_l + L_2) - (H_r - L_1)}{L_1 + L_2} \tag{1.11}$$

where L_1 and L_2 represent the length of the roots and plant stem, respectively (shown in Figure 1.10b), and H_l and H_r are the hydraulic head in plant leaves and roots, respectively (including osmotic suction and matric suction given by Eq. [1.1]). Note that not only matric suction but also osmotic suction can induce the hydraulic gradient necessary to drive water flow inside the plant. In Eq. (1.11), the water transport height $(L_1 + L_2)$ is mainly determined by the capillary force within the cell walls of the xylem (Nobel, 2009). The capillary height is mainly affected by the radius of capillary water in plants, which may be idealised as follows (Fredlund and Rahardjo, 1993):

$$h_c = \frac{2T_s \cos\beta}{\rho_w g R_s} \tag{1.12}$$

where h_c is the capillary height, T_s is the surface tension, R_s is the capillary radius, ρ_w is the water density, g is the force of gravity and β is the contact angle indicating the degree of soil hydrophilicity. The contact angle of the cell wall in plants is usually assumed to be 0° (Nobel, 2009). At 20°C, the capillary height can be expressed as

$$h_c = \frac{1.49 \times 10^{-5} (\text{m}^2)}{R_s} \tag{1.13}$$

where both h_c and R_s are measured in metres. As the radius of xylem vessels typically ranges from 8 to 500 μm, the corresponding range of capillary heights varies from 0.03 to 1.86 m, according to Eq. (1.13). Such small capillary heights can explain the water transport only in short plants. For taller plants (e.g., those that are 30 m tall, with an average xylem radius of about 0.5 μm), it is not possible to explain how water can be driven to such heights based on the capillary height generated by the numerous interstices in the cell wall of xylem vessels. It should be noted that the lumens of xylem vessels are not open to the atmosphere at the upper end. The numerous interstices in the cell wall of xylem vessels form a meshwork of many small, tortuous capillaries, which can drive water to great heights within a tree (Nobel, 2009). A representative 'radius' for these channels in the cell wall might be 5 nm (Nobel, 2009). With such a tiny radius, the capillary height can reach 3 km, according to Eq. (1.13). Other mechanisms are involved, such as the cohesion–tension mechanism, which refers to the phenomenon where hydrogen bonds are formed between water molecules along the path of water transport from roots towards leaves. The pressure inside the capillaries of trees can be up to 30 MPa (McElrone et al., 2013).

1.4.3 Repair of xylem cavitation

When negative pore water pressure in the xylem vessels of a plant is low, cavitation will occur to impede the transport of water inside the plant (Sperry et al., 1998). Cavitation occurs when air bubbles form within a vessel and block the transport of water from fine roots to leaves. The possibility of cavitation increases with a larger xylem diameter. This explains why the fine roots, and not the roots with a larger diameter, are mainly responsible for water uptake. To transport water to a greater height after cavitation, plants rely on certain biological mechanisms to overcome or repair the 'damage' caused by cavitation (Sperry et al., 1998). They decrease their transpiration by closing stomata on leaves. They also use their repairing system to reduce xylem cavitation (Figure 1.11) (Brodersen et al., 2010). When there are air bubbles inside the xylem vessels, embolism triggers the surrounding cells to secrete solutes into the vessels. The increased osmotic suction inside the xylem vessels can establish an osmotic gradient, drawing water from the surrounding fibres and parenchyma cells into the vessels. When the xylem vessels are successfully refilled, the refilling water would compress the air bubbles in the xylem, until they dissolve or escape into the surrounding hydrophobic microchannels in the vessel wall (Brodersen et al., 2010). The repaired xylem vessels can then transport water freely again. It is believed that the processes of cavitation and repair occur frequently and simultaneously to ensure that enough water can be transported from the roots to leaves for photosynthesis. However, solid scientific and biological proof is still needed to be explored.

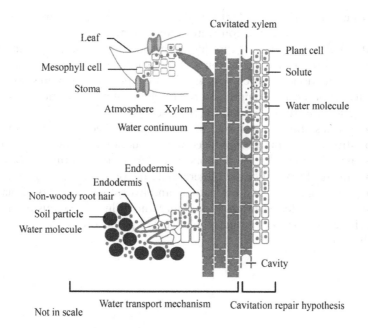

Figure 1.11 Water transportation from soil to the atmosphere via plants and an illustration of the repair mechanism of cavitation. (Revised from Brodersen, C. R., et al., *Plant Physiol.*, 154, 1088–1095, 2010.)

1.5 STRUCTURE OF THE BOOK

Plants interact strongly and dynamically with the soil, water and the atmosphere, and hence, they can affect the stability and deformation of soil slopes. This book presents our current understanding of mechanical and hydrological plant–soil slope interaction via an all-round, cross-disciplinary approach (see Figure 1.12). Chapter 2 describes a series of comprehensive

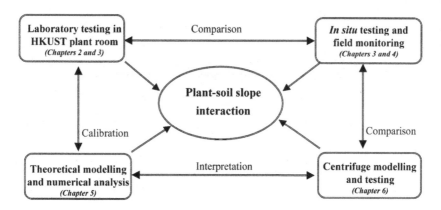

Figure 1.12 Layout of the book.

laboratory testing of the effects of plant transpiration on the induced matric suction and on soil hydraulic properties, including soil water retention and water permeability. Chapter 3 provides a state-of-the-art review of root biomechanical properties, including the root tensile strength and pull-out resistance of some of the plant species that are native to southern China but seldom mentioned in the literature. This chapter also explores how microorganisms (fungi) affect root tensile strength and the mechanical reinforcement of soil. Chapter 4 discusses three field case studies examining the effects of soil–plant–atmosphere interaction on the hydrology and stability of soil slopes. In Chapter 5, theoretical closed-form solutions are derived to estimate transpiration-induced matric suction and the stability of an infinite unsaturated slope reinforced by plant roots of different architectures. Transient seepage analysis on the effects of plant-induced preferential flow on matric suction is also presented. Finally, in Chapter 6, the combined mechanical and hydrological effects of plants of different root architectures on slope hydrology, stability and failure mechanisms are investigated through geotechnical centrifuge testing. The effects of plants on surface erosion are beyond the scope of this book.

Chapter 2

Hydrological effects of plant on matric suction

2.1 INTRODUCTION

To increase the shear strength and reduce the water permeability of soil, engineers will try to compact the soil to a sufficiently high density, for example, to a relative compaction (RC) of 95% (GEO, 2011). However, plants grow more effectively in a looser-state soil. These two contradictory requirements impose great challenges on the use of vegetation. This chapter focuses on two factors that can affect the design of soil bioengineering techniques. The first factor is soil density. In practice, it is not straightforward to determine an optimum compaction level of soil that can satisfy both ecological and engineering requirements. The second factor is plant density. Improved ecological restoration of civil engineering systems such as engineered slopes requires the knowledge of the optimum plant density. Knowing the optimum plant arrangement can help to minimise unfavourable competition among plants (Azam-Ali et al., 1984; Darawsheh et al., 2009). The optimum plant density can encourage plant growth and potentially maximise the beneficial effects of evapotranspiration (ET)-induced suction on the stability of these systems. As discussed in Section 1.3, carbon dioxide (CO_2) is essential for photosynthesis and hence plant growth. Recognising climate change and the potentially elevated concentration of atmospheric CO_2 is thus needed for improving our understanding on how the elevation of CO_2 concentration may affect transpiration-induced suction.

This chapter discusses the hydrological effects of several grass, shrub and tree species native to southern China and a temperate region of Europe on matric suction (MS) in sandy soil under different soil and plant densities. By establishing correlations with relevant above- and below-ground plant traits, the magnitude of suction induced by different plant types may be explained through plant ecology, anatomy and physiology. The effects of plant roots on soil water retention behaviour and water permeability are also discussed, which provide the necessary experimental evidence to support the hypothesis and modelling assumptions made in simulating plant hydrological effects in Chapter 5.

2.2 FACTORS CONTRIBUTING TO TRANSPIRATION-INDUCED SUCTION

2.2.1 Atmospherically controlled plant room

The Hong Kong University of Science and Technology (HKUST) is equipped with an atmospherically controlled plant room, as shown in Figure 2.1. The room can regulate an air temperature from 10°C to 30°C ± 1°C and an air relative humidity (RH) from 40% to 70% ± 5%. The room is also equipped with cool white fluorescent lamps, which emit light with a wavelength of 400–700 nm, favourable for plant photosynthesis and growth (Gates, 1980).

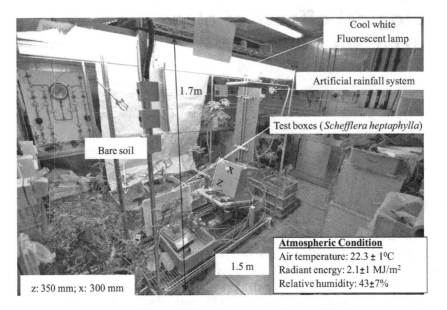

Figure 2.1 Overview of the atmospherically controlled plant room at the HKUST along with test boxes containing soil with different types of vegetation.

A quantum sensor (see working principle at Campbell Scientific Inc., 2008) is used to measure the radiant energy emitted from the lamps. By controlling the number of lamps, the amount of radiant energy can be controlled in the room to reach 15.3 ± 1 MJ/m²/d. According to the Penman equation (Penman, 1948), the potential evaporation (i.e., the maximum amount of water leaving the soil surface as vapour under abundant water supply) is about 4.5 mm/d.

In the atmospherically controlled plant room at the HKUST, it is possible to study not only the maximum amount of suction that can be induced by transpiration under drying conditions but also the minimum suction that can be preserved by plant roots during wetting, such as the use of a rainfall simulator shown in Figure 2.2. The room is carefully designed, so that most of the components of the water balance equation in Eq. (1.6) can be quantified.

Note that in this book, water storage in plant (ΔW_{plant}) is assumed to be negligible, as compared with the water retained in soil, i.e., ΔS_{soil} (McElrone et al., 2013). Indeed, during transpiration, 95% of water uptake by plant roots would be transpired (McElrone et al., 2013). On the other hand, surface water pond (ΔW_{pond}) can be neglected because the soil surface is made slightly inclined (<2°) for allowing surface water drainage.

To apply rainfall events with different intensities and durations (i.e., to control P in Eq. [1.6]) during a test, the rainfall simulation system installed in the plant room was used. This system consists of 10 plastic tubes, each 20 mm in diameter. Each tube is drilled with holes 1 mm in diameter to allow water to be discharged as rainfall. The simulator is mounted on two support stands at a specified height (Figure 2.3) above the soil surface. It is connected to a water reservoir with a capacity of 2 L for maintaining a constant difference in hydraulic head. By adjusting the height of the rainfall simulator relative to the water reservoir, a constant rainfall intensity can be maintained. A flow meter with an accuracy of ±0.5 L/h is attached to the device so that any desired water discharge rate (i.e., rainfall intensity) and discharge duration (i.e., rainfall duration) can be applied during testing. By controlling P and measuring surface runoff (R_{off}), the amount of water infiltration (R_{in})

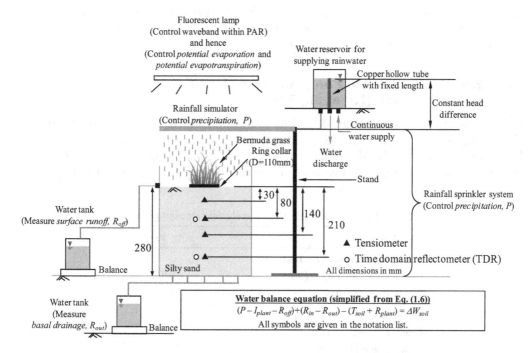

Figure 2.2 Schematic setup of a test box for studying the vegetation effects on soil water balance.

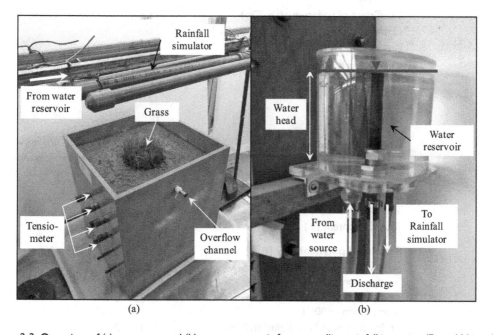

Figure 2.3 Overview of (a) test setup and (b) water reservoir for controlling rainfall intensity. (From Woon, E., Field and laboratory investigations of Bermuda grass induced suction and distribution, MPhil thesis, The Hong Kong University of Science and Technology, Hong Kong, 2013.)

is known. While water retained in soil (ΔW_{soil}) can be estimated from the measurements of soil moisture, the amount of root water uptake by plants (R_{plant}) can be determined via the water balance calculation given by Eq. (1.6).

2.2.2 Effects of soil density on plant growth and induced suction (Ng et al., 2014a)

To study the effects of soil density (RC of 70%, 80% and 95% of the maximum dry density of 1870 kg/m³) on ET-induced MS in grassed soil, six test boxes containing completely decomposed granite (CDG; silt sand; see index properties and nutrient content in Table 2.1) were prepared. These include three boxes of bare soil with different densities (test boxes

Table 2.1 Index properties of completely decomposed granite

Index properties	Value
Standard compaction tests	
Maximum dry density (kg/m³)	1870
Optimum moisture content (%)	12
Particle-size distribution	
Gravel content (>2 mm) (%)	19
Sand content (≤2 mm) (%)	42
Silt content (≤63 μm) (%)	27
Clay content (≤2 μm) (%)	12
D_{10} (mm)	0.15
D_{30} (mm)	0.7
D_{60} (mm)	2
Coefficient of uniformity (D_{60}/D_{10})	13.3
Coefficient of curvature ($[D_{30}]^2/[D_{60}D_{10}]$)	1.6
Specific gravity	2.60
Atterberg limit	
Plastic limit (%)	26
Liquid limit (%)	44
Plasticity index (%)	18
Chemical components	
Extractable N (mg/kg)	0.76
Extractable P (mg/kg)	0.53
Total C content (%)	1.2
Total K content (%)	7.0
Total Ca content (%)	0.96
Saturated permeability (m/s)	
At dry density of 1309 kg/m³ (relative compaction of 70%)	2.20×10^{-6}
At dry density of 1496 kg/m³ (relative compaction of 80%)	1.50×10^{-7}
At dry density of 1777 Kg/m³ (relative compaction of 95%)	1.10×10^{-9}
Unified Soil Classification System (USCS)	Silty sand (SM)

denoted by B70, B80 and B95) and three boxes of grassed soil (test boxes denoted by G70, G80 and G95). The grass tested was *Cynodon dactylon* (also known as Bermuda grass), which is a warm-season grass species widely cultivated in warm climates in many parts of Asia (Skerman and Riveros, 1990). This grass species has a high drought tolerance (Hau and Corlett, 2003) and is commonly used as a hydroseeding mix for non-engineered slope greening in Hong Kong. All six boxes were subjected to an artificial rainfall with a constant intensity of 100 ± 2 mm/h, using the rainfall simulator (Figures 2.2 and 2.3), for an hour, which corresponds to a return period of 100 years (Lam and Leung, 1995). During the rainfall, the responses of suction at four depths (i.e., 30, 80, 140 and 210 mm), amounts of surface water overflow R_{off} and basal percolation R_{out} (refer to Eq. [1.6]) in all six boxes were recorded. No fertiliser was added in any of the tests, thus preventing the different concentrations of solutes in pore water from inducing osmotic suction (Krahn and Fredlund, 1972; Ng and Menzies, 2007).

2.2.2.1 Grass characteristics

Figure 2.4 shows the measured average root depth varying with degree of compaction. It can be seen that the average root depth decreased by 36% as RC increased by 25%. A linear correlation was observed between root depth and RC. The measured smaller root depth in denser soil was attributed to increased mechanical resistance for the roots when penetrating soil pores (Bengough and Mullins, 1990). As roots penetrate the pore space in denser soil, soil particles are rearranged more substantially (Dorioz et al., 1993), which might create larger soil pores and allow roots to grow longer. In addition to mechanical resistance against root penetration, the root growth might be affected by reduced soil aeration at a high RC. Soil aeration contributes crucially to the root uptake of oxygen and is affected by air permeability, and hence the oxygen diffusion rate, in soil (Stepniewski et al., 1994; Granovsky and McCoy, 1997). The smaller pore size in denser soil is thus believed to lead to a reduced rate of oxygen diffusion, resulting in poorer root growth and smaller root depth. On the other hand, no clear correlation was identified between RC and grass shoot length.

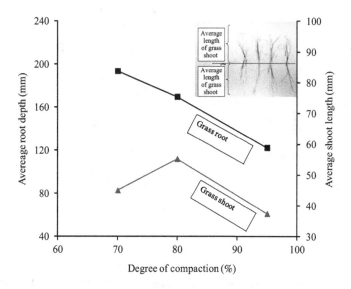

Figure 2.4 Effects of soil density on grass root depth and shoot length.

2.2.2.2 Water infiltration rate

During the tests, the infiltration rate of each test box could be determined by subtracting any measured surface water overflow, actual evaporation and actual evapotranspiration from the applied rainfall intensity (i.e., $P - R_{off} - T_{soil} - ET$; P is precipitation, R_{off} is surface runoff, T_{soil} is soil evaporation and ET is evapotranspiration); see Eq. (1.6). The terms T_{soil} and ET may be negligible compared with the applied rainfall intensity (P) of 100 mm/h. As shown in Figure 2.5, the infiltration rate in B70 was identical to the applied rainfall intensity for the entire rainfall event. This means that rainwater fully infiltrated the soil. For B80 and B95, the infiltration rates were equal to the rainfall intensity in the first 5 min but dropped afterwards. The infiltration rate in B95 was always lower than that in B80, because the water permeability of the soil compacted at a higher RC of 95% was nearly two orders of magnitude lower than that of the soil compacted at an RC of 80%. After 30 min of rainfall, the infiltration rates in both B80 and B95 appeared to reach a steady state. For vegetated boxes G70, G80 and G95, the infiltration was apparently not different from that in the corresponding bare soil.

2.2.2.3 Induced suction distribution

Figure 2.6 shows the distributions of MS in the six boxes during the rainfall event. After raining for 20 min, suction in the top 140 mm of soil in B70 dropped to 0 kPa but that at a depth of 210 mm remained unchanged at an initial value of about 35 kPa (Figure 2.6a). On the contrary, suction decreased to 0 kPa at all depths in G70 for the same rainfall duration (Figure 2.6b). Since the infiltration rates in B70 and G70 were similar (Figure 2.5), the observed greater depth of influence by rainfall in G70 means that the water permeability in G70 was higher than that in B70. At the end of rainfall, the wetting front in B70 advanced deeper and suction at the depth of 210 mm also dropped to 0 kPa.

For denser soil compacted at an RC of 80%, suction in the top 140 mm in B80 decreased substantially after 20 min of rainfall (Figure 2.6c). On the contrary, the depth of influence by rainfall in vegetated soil (G80) was shallower (Figure 2.6d) for the same rainfall duration.

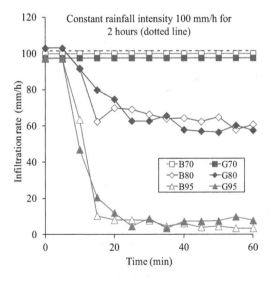

Figure 2.5 Effects of soil density on variations in infiltration rate with time in bare (B70, B80 and B95) and grass-vegetated (G70, G80 and G95) test boxes during rainfall.

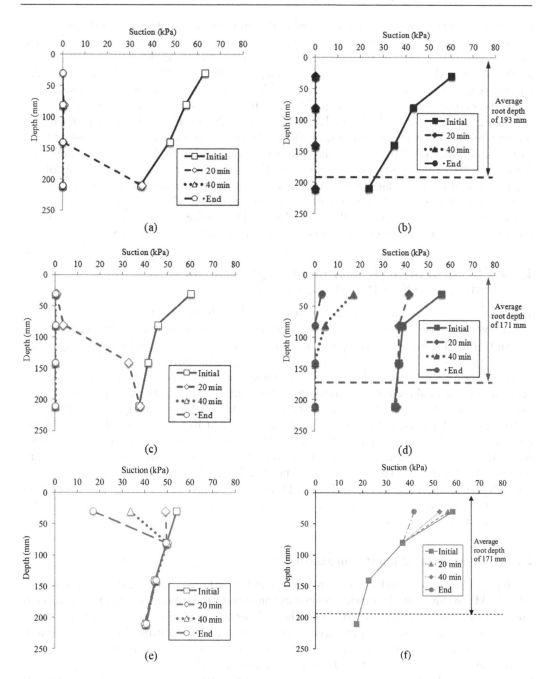

Figure 2.6 Suction distributions measured during a 60-min rainfall event in (a) B70, (b) G70, (c) B80; (d) G80, (e) G95 and (f) G95.

This indicates that the water permeability in the vegetated soil compacted at an RC of 80% was lower than that in the bare soil. At the end of the rainfall, suction recorded at all depths in B80 dropped to 0 kPa (Figure 2.6c), whereas a small amount of suction (i.e., 3 kPa) was retained at a depth of 30 mm in G80 (Figure 2.6d). For the densest compacted soil at an RC of 95%, only the suction at a depth of 30 mm showed a substantial drop in B95, whereas negligible changes in suction were recorded at the other three (greater) depths (Figure 2.6e). In contrast, no significant change in suction was recorded in the first 40 min of rainfall (see Figure 2.6f) at all depths. At the end of rainfall, suction along the depth was retained, except at 30 mm below the surface, where there was a drop of about 11 kPa, from 53 kPa to 42 kPa, during the last 20 min of rainfall. The observed higher suction retention capability at an RC of 95% was consistent with the measured lower infiltration rate shown in Figure 2.5.

Although looser soil at the lower RCs in G70 and G80 encouraged deeper root growth (hence a deeper root zone), any beneficial effects of roots on the magnitude of suction retained were less pronounced than those in the denser vegetated soil in G95. Among the three RCs considered, the presence of grass roots in G95 led to the lowest infiltration rate (8% of the applied rainfall intensity; Figure 2.5), the largest amount of suction preserved (up to 60% of initial suction before rainfall; Figure 2.6) and also the shallowest depth of influence by rainfall (about half of the average root depth; Figure 2.6e and f). In other words, the influence of grass on preserving suction during rainfall was minimal at an RC of 80% or looser.

2.2.3 Effects of plant density on plant growth and induced suction (Ng et al., 2016e)

This series of experiments is aimed at studying the effects of the density of *Schefflera heptaphylla* on tree growth and its influence on the induced suction in the CDG under *ET* and rainfall conditions. *Schefflera heptaphylla* is also known as the Ivy tree, which has sharp leaves and is common in many parts of Asia, including southern China, Japan, Vietnam and India (Hau and Corlett, 2003). This species has a significant ornamental and ecological value in slope rehabilitation and reforestation (GEO, 2011) and is also drought tolerant (Hau and Corlett, 2003). Ten tests were conducted – nine for vegetated soil having different plant densities and one for bare soil (control; test B) in the atmosphere-controlled room at the HKUST. Three plant densities, 320, 81 and 36 tree seedlings/m², were tested, corresponding to the plant spacings of 60 mm (test S60), 120 mm (test S120) and 180 mm (test S180), respectively. To take into account plant variability, three replicates were tested for each plant density. Hence, a total of 61, 25 and 13 seedlings of *S. heptaphylla* were transplanted to test drums with uniform tree spacings of 60 mm (S60), 120 mm (S120) and 180 mm (S180; Figure 2.7), respectively.

2.2.3.1 Above-ground plant characteristics

During the 4-month growth period, any change in the leaf area index (LAI) was monitored, where LAI is a dimensionless index defined as the ratio of the total leaf area to the projected area of canopy of an individual plant on the soil surface in the horizontal plane (Watson, 1947). Figure 2.8 shows the measured variation in tree LAI during the growth period. *Schefflera heptaphylla* grew with time. The seedlings grown at a lower plant density experienced a larger increase in LAI than the seedlings grown at a higher density, possibly because the lateral growth of leaves under a higher plant density was suppressed to a greater extent, as more leaves were shaded by adjacent plants. Shading reduced photosynthesis and hence leaf growth. Table 2.2 shows the range, mean and standard deviation of the above- and below-ground properties of all selected tree seedlings used in the nine vegetated soil tests.

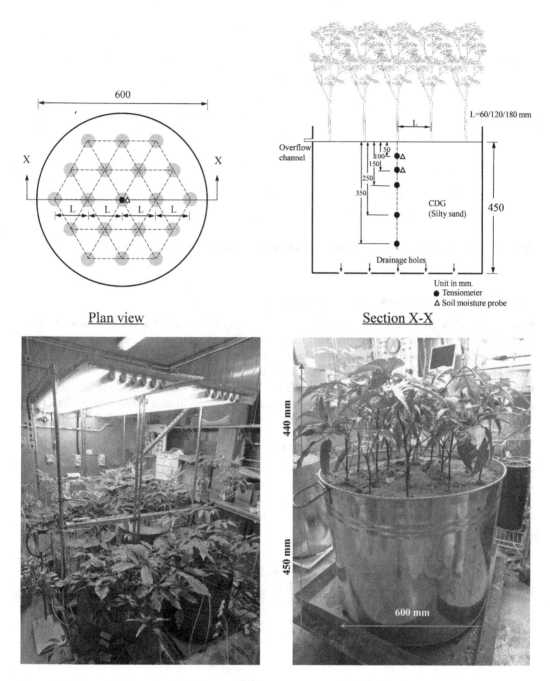

Plan view   Section X-X

Figure 2.7 Schematic diagrams and pictures of the test setup and instrumentation for studying effects of plant density in the plant room at the HKUST.

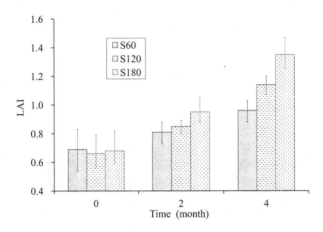

Figure 2.8 Changes in the LAI during the growth period under three different plant densities.

2.2.3.2 Below-ground plant characteristics

Root area index (RAI) of all tree seedlings was determined after testing, where RAI is defined as the ratio of the total root surface area to the circular cross-sectional area of soil in the horizontal plane for a given depth range (Francour and Semroud, 1992). The circular cross-sectional area of soil refers to the circular area whose diameter is defined by the maximum lateral spread of the root system within a given depth range. The root surface area refers to the total outside (external) surface area of all roots within a given soil volume that is defined by the cross-sectional area and the depth range. Detailed measurement methods of RAI can be found in Ng et al. (2016e).

Figure 2.9 shows that the distributions of RAI were non-linear and parabolic in shape in all cases. The RAI peaked at the depths of 60–90 mm. Interestingly, even though seedlings in test S60 had smaller root volumes than those in test S180 (to be discussed in Figure 2.10), the peak RAI within the top 10–90 mm of soil was around 34% higher than those of the other two cases (Figure 2.9). Figure 2.10 compares the root geometries of some of the tree seedlings

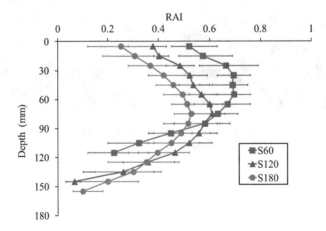

Figure 2.9 Effects of plant density on the RAI after a 4-month growth period.

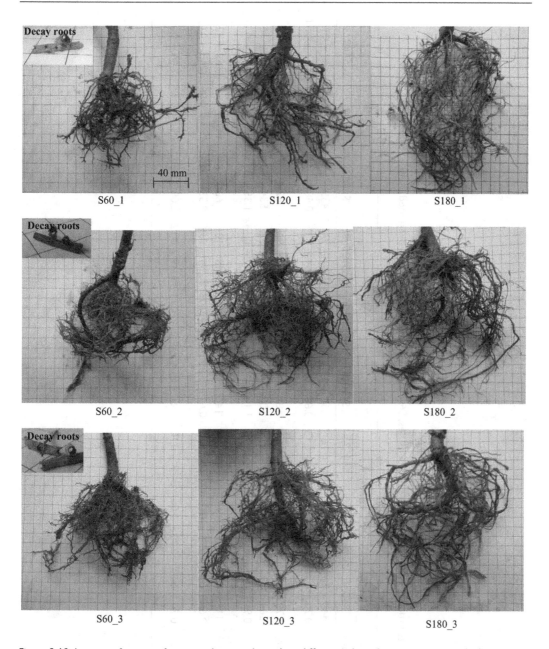

Figure 2.10 Images of roots of tree replicates planted at different plant densities excavated after testing (the sheet of paper in the background contains 10 × 10 mm² squares).

grown at the three plant densities after testing. In general, the roots of tree seedlings from test S60 were shorter and more localised. Some of the roots had decayed. On the contrary, the roots of seedlings from test S180 were longer and more dispersed. The roots were mainly fresh and were 17%–36% longer than those in test S60 (Table 2.2). With little lateral and vertical extension, the root volume in test S60 was only one-third of that in test S180.

At a higher plant density, the demand and competition for water are greater among neighbouring seedlings. A greater consumption and depletion of soil moisture would lead to a

Table 2.2 Summary of tree characteristics considering different plant densities, S60, S120 and S180 (all results are presented in mean value ± standard deviation)

	Tree characteristics	S60			S120			S180		
		S60_1	S60_2	S60_3	S120_1	S120_2	S120_3	S180_1	S180_2	S180_3
After transplantation	Height (mm)	440 ± 10	436 ± 16	449 ± 26	438 ± 15	443 ± 21	443 ± 23	435 ± 13	445 ± 16	442 ± 27
	Leaf area index (LAI)	0.70 ± 0.10	0.68 ± 0.06	0.71 ± 0.13	0.68 ± 0.14	0.65 ± 0.14	0.64 ± 0.15	0.68 ± 0.09	0.66 ± 0.10	0.72 ± 0.11
	Root depth (mm)	78 ± 10	84 ± 8	74 ± 16	81 ± 12	85 ± 11	84 ± 9	82 ± 9	80 ± 13	87 ± 12
Four months after growth	Height (mm)	564 ± 39	583 ± 68	586 ± 32	600 ± 48	544 ± 39	572 ± 52	584 ± 63	575 ± 24	560 ± 36
	Leaf area index (LAI)	0.91 ± 0.05	1.02 ± 0.09	0.95 ± 0.10	1.10 ± 0.04	1.12 ± 0.08	1.20 ± 0.14	1.35 ± 0.11	1.26 ± 0.09	1.40 ± 0.11
	Root depth (mm)	118 ± 10	126 ± 14	137 ± 13	146 ± 13	145 ± 12	148 ± 10	160 ± 18	147 ± 19	165 ± 12
	Peak root area index (RAI)	0.66 ± 0.10	0.75 ± 0.04	0.68 ± 0.09	0.61 ± 0.05	0.65 ± 0.05	0.54 ± 0.06	0.53 ± 0.07	0.60 ± 0.06	0.48 ± 0.09
	Root volume ($\times 10^6$ mm^3)	0.74 ± 0.07	0.70 ± 0.06	0.79 ± 0.10	1.41 ± 0.06	1.38 ± 0.11	1.44 ± 0.12	2.08 ± 0.08	2.12 ± 0.13	2.04 ± 0.12

reduction in root activity (Casper and Jackson, 1997; Jiang et al., 2013). This explains why the root growth for all seedlings was much more localised in test S60 than that in test S180, where seedling competition was less intense. However, owing to the competition of water among trees, plants would generally activate abscisic acid for root proliferation (hence RAI increases; Figure 2.9) and survival (Munns and Sharp, 1993).

2.2.3.3 Suction induced during evapotranspiration

During tests, the soil surface of all test drums was exposed to identical constant atmospheric conditions in the plant room for evaporation (test B) and *ET* (tests S60, S120 and S180) for 4 days. Figure 2.11 shows that the MS induced by vegetation increased much more significantly, regardless of the plant density, than that in the bare soil. The peak *ET*-induced suction at a depth of 50 mm in the vegetated soil was 64%–425% higher than that in the bare soil. Tree seedlings spaced closer to each other induced higher suction at all depths for a given duration of *ET*, owing to more intense competition among trees for water. Within the root zone, the amount of suction induced at a depth of 50 mm was always greater than that induced at a depth of 100 mm. In addition to the effects of surface evaporation, a major reason for higher suction was the higher values of RAI at shallower depths (Figure 2.9), where root water uptake was likely to be greater. Even though the roots of the tree seedlings in test S60 were the shallowest among the three cases, they exhibited the deepest influence zone of induced suction.

Figure 2.12 shows the correlation between *ET*-induced suction by *S. heptaphylla* in the root zone and LAI (Ng et al., 2016c, 2018a). To highlight the effects of tree transpiration, suction is expressed as the mean suction (*s*) difference between vegetated soil and bare soil, Δ*s*. It can be seen from the figure that Δ*s* has a strong correlation with LAI ($R^2 = 0.91$). A tree having a higher LAI intercepts more radiant energy, meaning that more stomata are available to absorb energy for transpiration (Kelliher et al., 1995). Correlations between Δ*s* and mean peak RAI are shown in Figure 2.13. As expected, for a given RAI, Δ*s* measured after ponding is always lower than that after *ET* because of infiltration. Δ*s* recorded either after ET or after ponding is strongly correlated with the RAI ($R^2 = 0.96$).

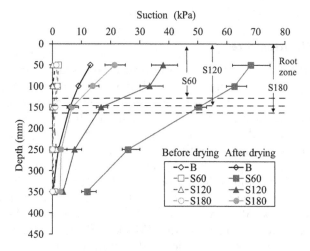

Figure 2.11 Comparison of the measured suction profiles of the bare and tree-vegetated soils after drying.

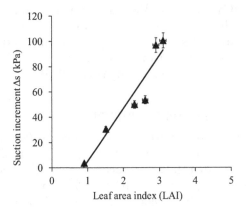

Figure 2.12 Relationship between evapotranspiration-induced suction increment by *S. heptaphylla* and LAI [$f = 41.9*x - 36.8$; $R^2 = 0.91$].

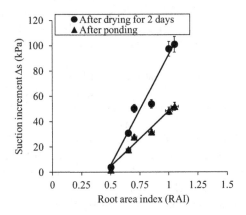

Figure 2.13 Relationship between suction increment by *S. heptaphylla* and RAI before and after ponding [$f = 173.3*x - 80.9$; $R^2 = 0.96$ for data obtained after drying for 2 days; $f = 86.7*x - 38.6$; $R^2 = 0.97$ for data obtained after ponding].

2.2.3.4 *Water infiltration rate*

Immediately after the 4-day evaporation and transpiration, all drums were subjected to a rainfall event simulated by using the rainfall device. A constant rainfall intensity of 73 mm/h was applied and maintained for 2 h, equivalent to a return period of 10 years in Hong Kong (Lam and Leung, 1995). The measured infiltration rates of bare soil and vegetated soil during rainfall are shown in Figure 2.14. As expected, the infiltration rate for the bare soil decreased exponentially with time, approaching the steady-state condition and k_s of the CDG (i.e., 1.2×10^{-8} m/s). The measurements show that the infiltration rate for the vegetated soil could be higher or lower than that for the bare soil, depending on the plant density. When the tree spacing was 120 or 180 mm, the infiltration rate for the vegetated soil was 18%–58% lower than that for the bare soil. This trend appears to be consistent with the findings in past studies (Meek et al., 1992; Ng et al., 2014a; Leung et al., 2015b), which also showed a reduced infiltration rate for soil containing actively growing roots.

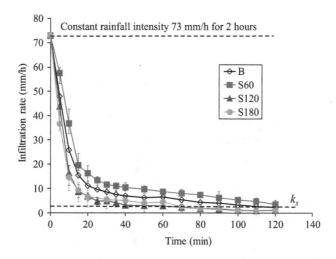

Figure 2.14 Variations in water infiltration rates with time for the bare and vegetated soils.

In contrast, when the trees were planted closer together (spacing = 60 mm), the infiltration rate for the vegetated soil was 42%–86% higher than that for the bare soil, likely because of the formation of preferential flow paths, as macro-pores were created in the soil due to the observed decay of roots (Ghestem et al., 2011). In addition to the effects of decayed roots, plant density might have also affected the infiltration rates because of the different stages of desaturation (i.e., due to transpiration-induced suction before rainfall occurred; Figure 2.11).

2.2.3.5 Suction preserved during rainfall

Figure 2.15a–d shows the vertical suction distribution before and right after the 2 h rainfall for the bare soil and the vegetated soils with the three different planting densities. After raining for 2 h, suction was reduced in the top 150 mm of the bare soil, and only minimal suction (i.e., <3 kPa) was preserved (Figure 2.15a). On the contrary, with roots in the soil in tests S120 and S180 (Figure 2.15b and c), rainfall mainly affected the suction within the root zone. This means that for a given period of rainfall, the depth of influence of suction in both vegetated soils was shallower than that in the bare soil. This observation is consistent with the test results shown in Figure 2.14 that the infiltration rates for the vegetated soils having the tree spacings of 120 and 180 mm were lower than that for the bare soil. For the tree spacing of 60 mm (Figure 2.15d), the influence zone of suction was deeper, as suction was found to change not only within the root zone but also at a depth of 150 mm. This might be attributed to the increased water permeability associated with root decay, as indicated by the increased infiltration rate in Figure 2.14. In contrast, when inspecting the suction responses below the root zone at a greater depth of 250 mm, the suction measured during the rainfall event was largely unchanged and preserved in all cases. The highest suction in the S60 samples (~26 kPa) was the consequence of the more significant tree root water uptake due to the smaller plant spacing (see Figure 2.11) before the rainfall occurred.

During the relatively short period of applied rainfall, *ET* of the vegetated soil or evaporation of the bare soil was minimal compared with the amount of rainwater that had infiltrated the soil. Thus, the different initial suctions before rainfall and hence different

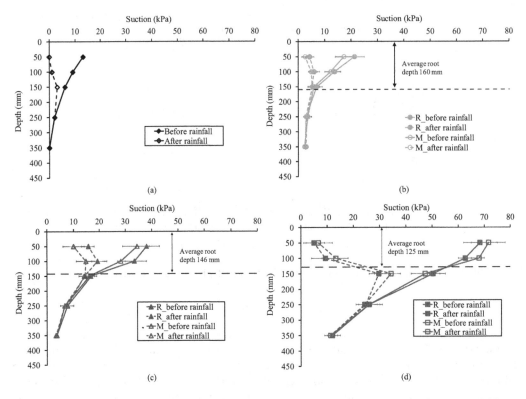

Figure 2.15 Measured suction profiles before and after rainfall: (a) test B, (b) test S180, (c) test S120 and (d) test S60.

water permeability (Ng and Menzies, 2007) might have been the major factors driving the suction responses during the subsequent rainfall. Figure 2.16a shows the measured suction within the root zone (i.e., 50 and 100 mm) before and after rainfall in all cases. For tree seedlings planted at the wider spacings of 120 and 180 mm, the amount of suction preserved after rainfall is positively correlated with the amount of initial suction before rainfall. The gradient of the correlation does not fully follow the 1:1 line, because not all suction was preserved after rainfall. Interestingly, the data obtained from test S60 did not follow the same trend and deviated from the 1:1 line. Although the magnitude of *ET*-induced suction before rainfall was the highest (Figure 2.11), the trees in this test preserved relatively little suction after rainfall (Figure 2.15a). This suggests that for this particular tree species grown in CDG, the benefit of the additional suction induced by *ET* within the root zone would be lost if the plant density was too high, as the resulting competition among trees might cause root decay and hence an increased rate of infiltration (Figure 2.14).

Figure 2.16b shows the relationships between suction before and suction after rainfall at the greater depths of 150, 250 and 350 mm. In most cases, except for the S60 samples at a depth of 150 mm, the suction induced by *ET* before rainfall was largely preserved after rainfall, as the data almost lie on the 1:1 line. Although suction was not preserved in all S60 samples, the amount of suction after rainfall was the highest. Interestingly, this observation is the opposite of the suction responses found within the root zone, where relatively little suction was preserved in the S60 samples (Figure 2.15a). This highlights that the higher the

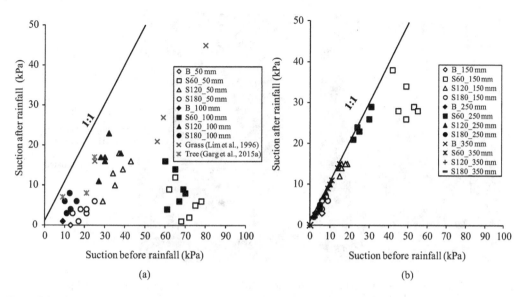

Figure 2.16 Correlations of measured suction before and after rainfall at depths of (a) 50 and 100 mm (From Lim, T.T. et al., *Can. Geotech. J.*, 33, 618–628, 1996) and (b) 150, 250 and 350 mm.

transpiration-induced suction before rainfall, the greater the influence on suction by plants to a deeper soil depth and hence the greater the amount of suction preserved at depth.

2.2.4 Effects of CO_2 on plant growth and induced suction (Ng et al., 2018b)

The same CDG and *S. heptaphylla* adopted in Section 2.2.3 were used to study the effects of atmospheric CO_2 and nutrient on induced soil suction. Three series of tests were conducted. The first series was to investigate the effects of two different concentrations of CO_2 of 400 (i.e., current atmospheric CO_2 level) and 1000 ppm (which is estimated to be the CO_2 concentration in year 2100) (IPCC, 2013) on plant growth in nutrient-deficit CDG (denoted as P400 and P1000). It should be noted that the current level of CO_2 concentration is about 400 ppm. The second series of tests was to study the influence of the same two CO_2 concentrations but in nutrient-supplied CDG (denoted as PN400 and PN1000), using a water-soluble fertiliser in which nitrogen (N), phosphate (P) and potassium (K) were provided in a ratio of 30%:10%:10% by volume in the fertiliser. The remaining 50% was composed of inert matter (Ng et al., 2018b). The last series of tests was on bare soil without vegetation, denoted as B. Three replicates were used for each test. All tests were conducted in an atmospherically controlled growth chamber (Figure 2.17). The inside wall of the chamber was coated with an aluminium fabric, so that the environment inside and outside of the chamber would not interact with each other. The setup was similar to the HKUST plant room (Figure 2.1), except that the chamber was connected to a compressed CO_2 tank for supplying and controlling different concentrations of CO_2 within the chamber. Since CO_2 is heavier than air, a rotating fan was placed near the CO_2 supply pipe to distribute CO_2 more uniformly within the chamber. The chamber was sealed to prevent loss of CO_2 and to maintain an identical atmospheric condition throughout all tests. The air temperature and RH in the chamber were regulated at 28°C ± 1°C and 60% ± 5% inside the sealed chamber, respectively.

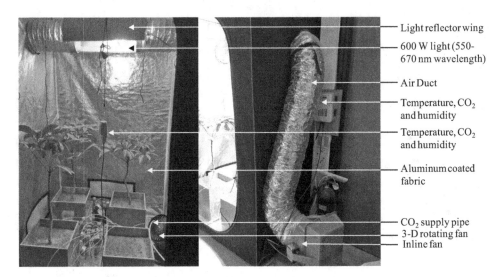

Figure 2.17 Overview of the test setup and instrumentation in the atmospheric-controlled plant growth chamber at the HKUST.

2.2.4.1 Plant characteristics

Figure 2.18a compares the LAI of *S. heptaphylla* during the 3 months of growing period under 400 and 1000 ppm CO_2. When the nutrient was supplied, there was a significant increase in the average LAI under higher CO_2 concentration of 1000 ppm (i.e., PN1000). At the last 30 days of growth, the rate of increase in LAI was almost the same in PN400 and PN1000 tests. On the contrary, for the CDG without supplying nutrient, the average LAI increased noticeably only under 400 ppm CO_2 (P400); however, it remained apparently constant under the elevated CO_2 concentration of 1000 ppm (P1000). Figure 2.18b compares

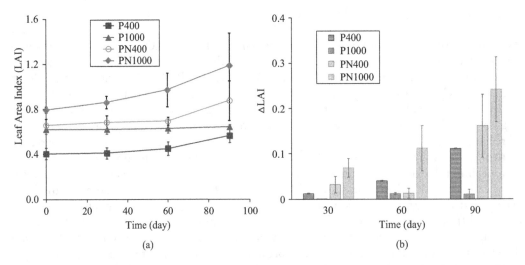

Figure 2.18 (a) Comparison of measured variation of LAI with time during 3 months of growing period and (b) changes of LAI after 30, 60 and 90 days of growth under four different treatments. Data points are presented as mean value ± standard error of mean ($n = 3$).

the LAI change (ΔLAI) before and after 90 days of growth. Evidently, as the atmospheric CO_2 concentration increased from 400 to 1000 ppm, the LAI of the tested species grown in nutrient-deficit soil reduced significantly (refer to P400 and P1000). The scarcity of soil nitrogen might have limited the plant response to elevated CO_2 concentration and hence leaf growth (Reich et al., 2006; Langley and Megonigal, 2010). Indeed, plant growth requires nitrogen uptake from soil (Ng'etich et al., 2013). When sufficient nutrient was supplied, the plant growth could be improved significantly under the elevated CO_2 concentration (compare PN400 and PN1000). In other words, sufficient amount of nutrient must be supplied to ensure proper plant growth under an increasing CO_2 concentration.

2.2.4.2 Induced matric suction

Figure 2.19 compares the suction profiles along depth before and after 6 days of plant transpiration. After 6 days of transpiration, non-uniform suction distribution was found in all vegetated soils owing to the parabolic shape of RAI distribution. As expected, vegetated soil induced higher suction than bare soil because of root water uptake, regardless of any CO_2 effects. When there was a lack of nutrient supplied in the soil, the amount of suction induced was lower under the elevated CO_2 concentration (Figure 2.19a). Indeed, Lewis et al. (2002) and Dong et al. (2002) also found that elevating ambient CO_2 concentration by 200 and 350 ppm in the atmosphere reduced transpiration rate by 12% and 53%–63%, respectively. During transpiration, plants lose water through open stomata, where CO_2 is absorbed simultaneously. A principal response of a plant to elevated atmospheric CO_2 is to reduce transpired water by reducing stomatal conductance, if no extra nutrient is supplied to promote its growth. A doubling of today's CO_2 levels from 390 to 800 ppm will halve the amount of water lost to the atmosphere (de Boer et al., 2011; Lammertsma et al., 2011). Thus, plant-induced suction decreased during transpiration under the elevated CO_2 concentration.

When nutrient was supplied to the soil (Figure 2.19b), higher suction was induced under both CO_2 concentrations considered. However, the effects of CO_2 on the magnitude and

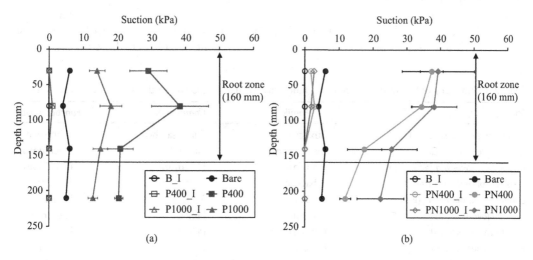

Figure 2.19 Measured suction profiles of (a) bare and nutrient-deficient vegetated soils and (b) bare and nutrient-supplied vegetated soils under 400 and 1000 ppm atmospheric CO_2 before and after 6 days of transpiration. Note that the initial suction profile in each sample is denoted with a letter 'I' at the end of the respective legend. Data points are presented as mean value ± standard error of mean ($n = 3$).

distribution of suction became less significant. The observed suction responses for the four conditions (P400, P1000, PN400 and PN1000) are consistent with the measurements of LAI reported in Figure 2.18. The higher the LAI, the higher would be the amount of suction.

2.3 CORRELATING PLANT TRAITS WITH INDUCED SOIL SUCTION (BOLDRIN ET AL., 2017B)

A plant trait is defined as a distinct and quantitative feature of a species in terms of plant morphology, physiology or biomechanics (Pérez-Harguindeguy et al., 2013). In order to better understand the soil—root—water interaction, it is important to investigate the correlation between plant traits and plant-induced suction. Indeed, there has been an increasing focus on using plant traits as screening criteria to assist engineers in identifying suitable species for slope stabilisation (Stokes et al., 2009). This section presents glasshouse tests to compare the hydrological reinforcement induced by 10 selected woody species widespread in Europe and to associate such reinforcement with functional traits corresponding to hydrological strategies and morphological characteristics.

Ten woody species, including *Buxus sempervirens* L., *Corylus avellana* L., *Crataegus monogyna* Jacq., *Cytisus scoparius* (L.) *Link*, *Euonymus europaeus* L., *Ilex aquifolium* L., *Ligustrum vulgare* L., *Prunus spinosa* L., *Salix viminalis* L. and *Ulex europaeus* L., were tested. Their family, common name, functional type and acronym are summarised in Table 2.3. These shrub and small tree species are widespread in Europe and have relatively high adaptability to diverse environmental conditions. Most of these species are within the Trunk Road Biodiversity Action Plan recommended by the Scottish Government for enhancing the ecological value and landscaping of roadside slopes and/or embankments. These species have been suggested as suitable plants for soil bioengineering and eco-technological solutions in the European context (Coppin and Richards, 1990; Marriott et al., 2001; Norris et al., 2008; Beikircher et al., 2010). In particular, *C. avellana* and *S. viminalis* are found to be highly suitable for slope stabilisation through mechanical reinforcement (Bischetti et al., 2005; Mickovski et al., 2009).

All these selected species were planted in the soil collected from Bullionfield, The James Hutton Institute, Dundee, U.K., at an initial dry density of 1200 kg/m^3 to facilitate fast root growth and development during plant establishment (Loades et al., 2013). The soil

Table 2.3 Summary of family, common name, functional type and acronym of the 10 selected test species

Species	Family	Common name	Functional type
Buxus sempervirens L. (Bs)	Buxaceae	European box	Evergreen
Corylus avellana L. (Ca)	Betulaceae	Hazel	Deciduous
Crataegus monogyna Jacq. (Cm)	Rosaceae	Hawthorn	Deciduous
Cytisus scoparius (L.) *Link* (Cs)	Fabaceae	Scotch broom	Evergreen
Euonymus europaeus L. (Ee)	Celastraceae	Spindle	Deciduous
Ilex aquifolium L. (Ia)	Aquifoliaceae	Holly	Evergreen
Ligustrum vulgare L. (Lv)	Oleaceae	Privet	Deciduous
Prunus spinosa L. (Ps)	Rosaceae	Blackthorn	Deciduous
Salix viminalis L. (Sv)*	Salicaceae	Willow	Deciduous
Ulex europaeus L. (Ue)	Fabaceae	Gorse	Evergreen

Source: Boldrin, D. et al., *Plant Soil*, 2017b.

*Indicates propagation by cutting. All plants were supplied by British Hardwood Tree Nursery, Gainsborough, UK.

was a sandy loam, which comprised 71% sand, 19% silt and 10% clay contents. More index properties of the soil can be found in Boldrin et al. (2017a, 2017b, 2018). Five replicates of each species were prepared, giving a total of 50 potted plants. The top soil surface of the pots was covered with a 10-mm-thick gravel layer to minimise evaporation. Three pots of fallow soil were prepared as control.

After initial plant establishment, all 50 potted plants and the pots containing fallow soil were left in the glasshouse for ET and evaporation, respectively, for 13 days. Measured daily, water loss was assumed to be equal to the daily ET from the potted plants and the daily evaporation from the fallow soil. Daily transpiration from each potted plant was estimated from the difference between ET and evaporation in the period between days 2 and 9.

2.3.1 Plant traits and physiological responses

A number of plant traits were measured to shed light on the hydrological reinforcement induced by the 10 different species. The measured above-ground traits included the following:

- Specific leaf area (SLA; m²/kg), defined as the one-sided area of a fresh leaf divided by its oven-dry mass, expressed in m²/kg.
- Wood and leaf biomass (g) (i.e., green and non-green biomass). Note that for *C. scoparius* and *U. europaeus*, it was not possible to separate green from non-green biomass, because of the presence of partially green shoots and thorns. Therefore, only the total above-ground biomass was measured.
- Green mass ratio (GMR), defined as the ratio between green biomass and the total above-ground biomass (expressed in g/g).
- Plant height (PH; expressed in cm).
- Wood density (main stem; WD; expressed in g/cm³).

The below-ground traits included the following:

- Total root length (expressed in m).
- Root biomass (RB; expressed in g).
- Specific root length (SRL; root length by mass, expressed in m/g). Note that thick roots (>5 mm in diameter), if any, were processed and analysed separately to avoid overestimation of the root length.
- Root length density (RLD), defined as the total root length by the soil volume in the pots (expressed in cm/cm³)
- Root-to-shoot ratio (RSR), defined as the ratio of below-ground to above-ground biomass (expressed in g/g).

All plant traits were measured according to the standardised methodology proposed by Pérez-Harguindeguy et al. (2013). Ideally, LAI and RAI should also have been investigated.

Plant physiological response, that is, leaf conductance to water vapour (g_L; mmol m²/s), was measured on at least one leaf for all replicates by using a portable porometer. Measurements of g_L were made on a sunny day, when all potted plants showed an evident and stable water loss.

2.3.2 Relationships between plant traits and induced suction

The main above- and below-ground traits (Table 2.4) show significant differences among species. A principal-component (PC) biplot (Figure 2.20) shows that from the projection of plant traits and soil hydro-mechanical characteristics on the plane composed of the first

Table 2.4 Main above- and below-ground traits of each species (mean ± standard error of mean)

Species	Wood biomass (g)	Leaf biomass (g)	Specific leaf area (m²/kg)	Leaf conductance to water vapour (mmol m²/s)	Root biomass (g)	Specific root length (m²/g)
Buxus sempervirens L. (Bs)	12.79 ± 1.99	13.02 ± 2.87	8.55 ± 1.53b	44.30 ± 9.9a	7.4 ± 1.0	18.84 ± 1.14a
Corylus avellana L. (Ca)	28.88 ± 2.92	12.46 ± 1.24	22.01 ± 1.09d	55.24 ± 8.7abc	29.7 ± 3.1	21.10 ± 1.88ab
Crataegus monogyna Jacq. (Cm)	18.34 ± 2.48	7.16 ± 0.63	15.98 ± 0.42c	140.80 ± 20.4bcd	15.0 ± 0.9	26.31 ± 6.07ab
Cytisus scoparius (L.) Link (Cs)	162.1 ± 9.9		–	26.52 ± 7.9a	16.8 ± 0.7	28.47 ± 2.57ab
Euonymus europaeus L. (Ee)	38.03 ± 5.66	14.92 ± 1.46	17.92 ± 0.38c	49.98 ± 10.2ab	23.9 ± 6.1	18.65 ± 0.86a
Ilex aquifolium L. (Ia)	6.39 ± 0.49	8.00 ± 0.90	4.27 ± 0.24a	26.64 ± 5.7a	2.4 ± 0.3	28.68 ± 3.70ab
Ligustrum vulgare L. (Lv)	29.35 ± 3.17	14.28 ± 2.48	14.56 ± 0.88c	63.30 ± 14.6abc	22.0 ± 4.1	15.55 ± 1.16a
Prunus spinosa L. (Ps)	13.65 ± 1.37	6.65 ± 1.09	23.44 ± 0.86d	153.20 ± 17.4cd	15.5 ± 0.9	36.23 ± 5.31bc
Salix viminalis L. (Sv)	25.09 ± 4.40	2.94 ± 0.21	21.88 ± 0.44d	417.60 ± 124.4d	15.9 ± 1.1	64.52 ± 9.03c
Ulex europaeus L. (Ue)	68.80 ± 4.34		–	56.40 ± 8.4abc	12.7 ± 1.3	28.91 ± 1.62ab

Source: Boldrin, D. et al., *Plant Soil*, 2017b.

Note: The letters next to the values in the specific leaf area, leaf conductance to water vapour and specific root length columns indicate significant differences among species, as tested using one-way ANOVA followed by Tukey test (leaf conductance to water vapour and specific root length data were log transformed). Total biomass (i.e., the wood, leaf and root biomass combined) among species showed significant differences (p-values < 0.001, one-way ANOVA of log-transformed data).

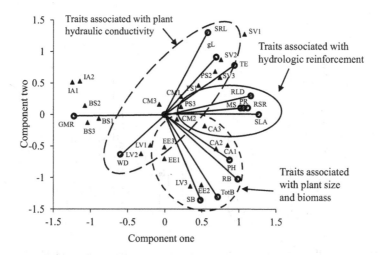

Figure 2.20 Biplot projection of plant traits and soil hydro-mechanical parameters on the plane, represented by the first two components of principal-component (PC) analysis (PC1: 48% of variation; PC2: 24% of variation). Acronyms of plant traits and soil parameters: g_L (leaf conductance), GMR (green mass ratio), MS (matric suction), PH (plant height), PR (penetration resistance), RB (root biomass), RLD (root length density), RSR (root-to-shoot ratio), SB (shoot biomass), SLA (specific leaf area), SRL (specific root length), TE (transpiration efficiency), Tot B (total biomass) and WD (wood density). (From Boldrin, D. et al., *Plant Soil*, 2017b.)

two explanatory axes (PC1: 48% of variation; PC2: 24% of variation), three major groups of plant traits can be defined. Since the vectors for the plant traits SLA, RLD and RSR have almost the same direction as the axis of component one (also known as first PC), these traits are positively correlated with each other. These traits are associated with soil hydro-mechanical reinforcement, including MS and penetration resistance (PR). On the other hand, the second PC axis is positively related to plant traits such as leaf conductance and SRL, which are associated with plant hydraulic conductivity (Eissenstat, 1992; Rieger and Litvin, 1999), and negatively related to traits such as plant height, shoot biomass, root biomass and total biomass, which are associated with plant size). The small angles in the biplot between soil hydro-mechanical characteristics and plant traits indicate that biomass allocation and investment (SLA, RLD and RSR) were strongly correlated with these parameters. On the contrary, plant traits associated with plant size were not correlated with soil hydro-mechanical characteristics. Leaf conductance, SRL and transpiration efficiency (transpiration per shoot biomass, expressed as g/g) were positively related to each other but negatively related to wood density.

It is generally recognised that plant water uptake is affected by biomass (both above- and below-ground), as well as physiological factors (Lambers et al., 2008; Osman and Barakbah, 2011; Jones, 2013). Interestingly, the PC biplot (Figure 2.20) shows that biomass allocation (e.g., the RSR) and biomass investment such as leaf surface area (e.g., SLA) and root length (e.g., RLD) were strongly and positively correlated with hydrological reinforcement (i.e., MS and PR). However, plant size and biomass were not correlated with MS (since the angle between these vectors is almost 90°), when the 10 different species were considered. The lack of correlation between biomass and water uptake in the experiment was also highlighted by the significantly different transpiration efficiency among species (Figure 2.21). Transpiration efficiency can be particularly relevant in species selection for soil hydrological reinforcement, as it is crucial to isolate the effects of biomass when estimating the effects of species on water

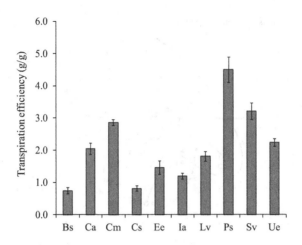

Figure 2.21 Transpiration efficiency (normalised daily transpiration per above-ground biomass). Means are reported ± standard error of mean (*n* = 5). Letters indicate significant differences among species, as tested using one-way analysis of variance (ANOVA), followed by Tukey test (data were log transformed). (From Boldrin, D. et al., *Plant Soil*, 2017b.)

uptake ability, so that the estimation is not biased by plant dimensions. This highlighted that other physiological factors differing among species, such as leaf conductance to water vapour, could have considerable effects on transpiration and transpiration efficiency, limiting the expected effects of biomass. In fact, transpiration efficiency was correlated with leaf conductance (g_L; Figure 2.22). For species such as *P. spinosa*, the high g_L may be one of the key factors compensating for the low biomass and inducing the relatively high suction.

The PC biplot shows strong correlations between hydrological reinforcement and certain plant traits (SLA, RLD and RSR), which may thus be used to identify the relative transpiration-induced suction from different species and the associated gain in soil strength.

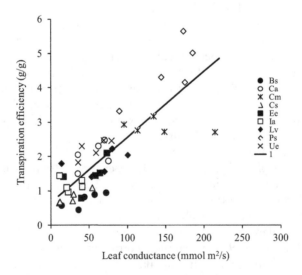

Figure 2.22 Relationship between leaf conductance to water vapour (g_L) and transpiration efficiency (daily transpiration per above-ground biomass) [$f = 0.6546 + 0.0191^*x$; *p*-value < 0.0001; $R^2 = 0.67$]. (From Boldrin, D. et al., *Plant Soil*, 2017b.)

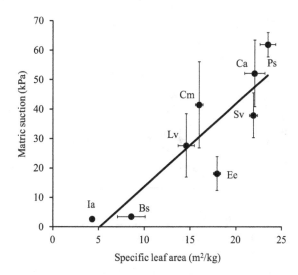

Figure 2.23 Relationship of specific leaf area (SLA) with matric suction. Mean values of species are reported ± standard error of mean ($n = 3$ for soil parameters; $n = 10$ for SLA). Linear regressions of all data points from all replicates (non-average values) are given [$f = -14.4182 + 2.8112*x$; p-value < 0.0001; $R^2 = 0.56$]. Note that *C. scoparius* and *U. europaeus* were not considered in the regression analyses because of the absence of leaves (*U. europaeus*) or their limited number and dimensions (*C. scoparius*) compared with green twigs and thorns, which are the main photosynthetic organs in these species. (From Boldrin, D. et al., *Plant Soil*, 2017b.)

For the above-ground traits, SLA showed a positive linear correlation with MS (Figure 2.23). Hence, it was not the leaf biomass but rather the biomass allocation and investment such as leaf surface area that determined MS. The SLA is an indicator of a species' energy strategy and adaptation to the environment. The SLAs of the selected deciduous species were higher than those of the selected evergreens (Table 2.4). The observed differences in SLA among the 10 species were attributable to the wide spectrum of leaf economics, which reflected the plant investment in leaf tissue (Wright et al., 2004). Thus, a low value of SLA implies more resistant leaves to grazing and mechanical damage and consequently a relatively longer leaf life span and a slow return on initial energy investment for the leaf (Wright et al., 2004; Poorter et al., 2009). In contrast, a high SLA value implies a fast return on energy investment, which would result in higher rates of net photosynthesis (Reich et al., 1997), potential growth (Grime et al., 1997) and transpiration (Reich et al., 1997). The fast return on energy investment represents the main biological reason for the correlation between SLA and MS (Figure 2.23). Under the temperate climate conditions in Europe, deciduous species are generally characterised by a high SLA and hence a faster return on energy investment and transpiration during the summer growing season (Bai et al., 2015). A recent study by Bochet and García-Fayos (2015) showed that SLA was a relevant trait for indicating plant competitiveness and success in establishing road embankments in semi-arid environments. Thus, SLA, whose measurement is relatively simple and quick, is a plant trait that can be used to assess the relative hydrological reinforcement and survival under the harsh environment of engineered slopes.

Among the below-ground traits, RLD showed a significant correlation with MS (Figure 2.24). Osman and Barakbah (2006, 2011) identified RLD as a relevant trait for both mechanical and hydrological reinforcement of soil. They found that RLD was positively

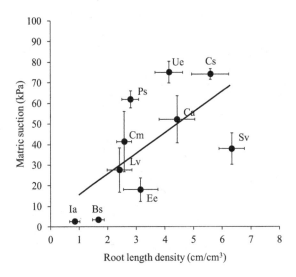

Figure 2.24 Relationship of root length density (RLD) with matric suction. Mean values of species are reported ± standard error of mean ($n = 3$). Linear regressions of all data points from all replicates (non-average values) are given [$f = -0.9510 + 13.2804*x$; p-value < 0.0001; $R^2 = 0.47$]. *Salix viminalis* was not included in the regression analysis because this species was grown from cuttings, whilst all other species were grown from seeds. (From Boldrin, D. et al., *Plant Soil*, 2017b.)

correlated with soil shear strength but negatively related to soil water content. In terms of the mechanical reinforcement, a high RLD means a higher cross-sectional area of roots traversing a potential shear surface per unit soil surface area (Ghestem et al., 2014). For hydrological reinforcement, RLD alone may not be sufficient to explain the amount of soil water depleted by a plant; however, a significant correlation was found. Other factors that could affect plant water uptake include a combination of other root traits such as the maximum root depth and specific root water uptake (Hamblin and Tennant, 1987).

For most of the plants studied, plant water uptake is not necessarily related to above- or below-ground traits only but to their ratio. The RSR showed the strongest correlation with MS (Figure 2.25) among all plant traits, except in *C. scoparius* and *U. europaeus*. This highlights the importance of considering the combined influence of above- and below-ground organs on the hydraulic effects induced by plants. It is hypothesised that the distinctive behaviour of *C. scoparius* and *U. europaeus* may result from their distinct photosynthetic twigs and thorns. Although their photosynthetic organs are photosynthetically analogues to leaves, the two species have a greater mass per surface area. Thus, *C. scoparius* and *U. europaeus* may require greater above-ground biomass investment to achieve the same photosynthetically active surface as broad-leaf species (i.e., the other eight species), resulting in a much larger shoot weight (i.e., low RSR).

Plant water uptake is the result of eco-physiological interactions between above- and below-ground processes. Roots contribute to the overall plant water demand, and they also account for 50%–60% of the hydraulic resistance of the entire plant, which substantially limits the water transport in the soil–plant–air continuum (Tyree and Ewers, 1991). Plant shoots, when referring to leaves and stomata, control and regulate plant water relations because of the steep gradient in water potential between a leaf and the atmosphere in that continuum (Steudle, 2001; Jones, 2013). Although both roots and shoots are important for water uptake, test results in Figure 2.25 show that an increase in the RSR could increase

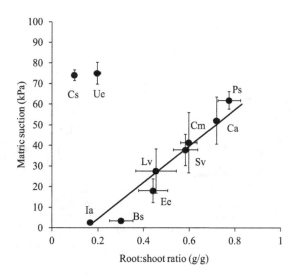

Figure 2.25 Relationship of the root-to-shoot ratio with matric suction resistance. Mean values of species are reported ± standard error of mean ($n = 3$). Linear regressions of all data points from all replicates (non-average values) are given [$f = -17.0648 + 93.3896*x$; p-value < 0.0001; $R^2 = 0.65$]. (From Boldrin, D. et al., *Plant Soil*, 2017b.)

hydrological reinforcement. The RSR may also be a relevant trait for mechanical reinforcement. Indeed, a higher RSR means that more roots are potentially contributing to mechanical soil reinforcement, whilst the above-ground biomass would be relatively small, inducing less surcharge and wind loading (Stokes et al., 2008) or seismic loading (Liang et al., 2015).

2.4 ROOT-INDUCED CHANGES IN SOIL HYDRAULIC PROPERTIES

It is generally recognised that the presence of plant roots in soil directly affects soil hydraulic properties, including soil water retention ability (Scanlan and Hinz, 2010; Ng et al., 2014a; Scholl et al., 2014; Leung et al., 2015a; Yan and Zhang, 2015; Jotisankasa and Sirirattanachat, 2017) and water permeability (Jotisankasa and Sirirattanachat, 2017), owing to the alteration of soil structures.

2.4.1 Water retention curve of vegetated soil

2.4.1.1 Soil vegetated with grass

Rahardjo et al. (2014) tested the effects of orange jasmine (*Myrrata exotica* L.) and vetiver roots (*Chrysopgon zizanioides*) on the water retention characteristics of old alluvium, with a fine content of about 20%–30%, using Tempe cells. Three major observations are shown in Figure 2.26. First, regardless of the types of grass species considered, the drying soil water retention curve (SWRC) of the vegetated soil was located above that of the bare soil. This indicates that the vegetated soil retains more water than the bare soil at a given suction. Second, the presence of roots in soil appeared to increase the soil air-entry value (AEV). Lastly, there was an evident change in the desorption rate (i.e., the drop in Volumetric water content (VWC) for a given increase in MS) owing to the presence of plant roots.

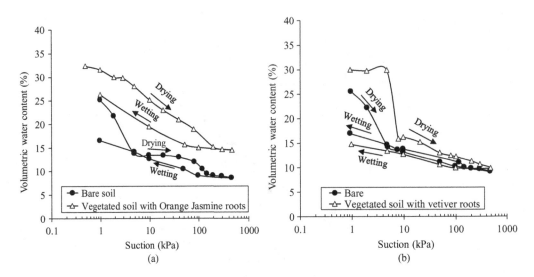

Figure 2.26 Comparisons of SWRCs of bare and grass-vegetated soils with (a) jasmine roots and (b) vetiver roots. (From Rahardjo, H. et al., *Soils Foundations*, 54, 417–425, 2014.)

Interestingly, Jasmine roots exhibited fairly constant desorption rate (Figure 2.26a), whilst vetiver roots led to an abrupt increase in the rate, between 5 and 9 kPa (Figure 2.26b).

Jotisankasa and Sirirattanachat (2017) adopted the instantaneous profile method (Watson, 1966; Ng and Leung, 2012; Leung et al., 2016) to investigate the effects of Vetiver roots (*C. zizanioides*) on the hydraulic properties of two types of soil, silt and clayey sand. Figure 2.27a shows the SWRCs of silt with various RLDs. As the RLD increased up to 6.5 and 7.14 kg/m³, the AEVs of the silt samples appeared to decrease slightly from respective 2.3 and 1.8 kPa to 0.7 kPa, probably owing to micro-cracks induced by wetting and drying cycles. Specimens with micro-cracks would desaturate at a lower suction than those without micro-cracks. Peng et al. (2007), Bodner et al. (2013) and Ma et al. (2015) also revealed an

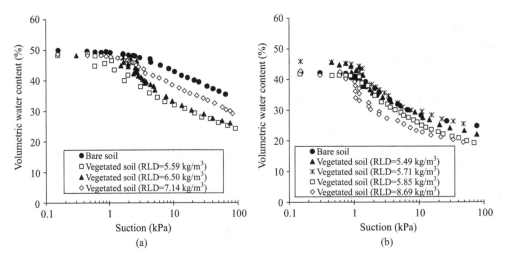

Figure 2.27 Comparisons of SWRCs of bare and grass-vegetated soils: (a) silt and (b) clayey sand. (From Jotisankasa, A. and Sirirattanachat, T., *Can. Geotech. J.*, 54, 1612–1622, 2017.)

increase in large pores, induced by wetting and drying cycles as a consequence of crack and micro-crack formation.

Figure 2.27b shows the SWRCs of the clayey sand with different RLDs. Except for case with RLD = 5.71 kg/m³, the VWC (θ_s) decreased with an increase in RLD at a given suction. As the RLD increased to 8.69 kg/m³, θ_s reduced, possibly owing to the more hydrophobic behaviour of the organic exudation from roots. When the RLD reached 8.69 kg/m³, the desaturation rate (i.e., reduction of θ_s due to an increase in suction) of the vegetated sample was steeper than that of the bare sample. This may be because of root decay hypothesis, which caused an increase in saturated permeability. The AEVs, nevertheless, appeared to be relatively unchanged as the RLD increased.

2.4.1.2 Soil vegetated with tree

Figure 2.28a shows the drying SWRC of a bare soil and a soil vegetated with *S. heptaphylla* at three different plant densities (test denoted as B, S60, S120 and S180; see Section 2.2.3). The SWRCs were obtained by relating the measured VWC to suction at a depth of 50 mm during the 4-day evaporation (for bare soil) and *ET* (for vegetated soil). The equation proposed by van Genuchten (1980) was used to fit the SWRCs. All required fitting parameters are shown in Table 2.5. The water retention ability of the rooted soil was different from

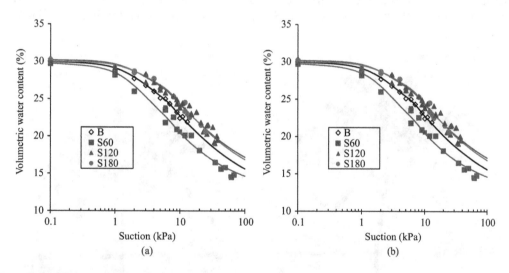

Figure 2.28 Effects of plant density on SWRCs of bare soil and tree-vegetated soil along (a) the drying path and (b) the wetting path.

Table 2.5 Summary of fitting parameters for van Genuchten (1980) equation

Test	Drying SWRC					Wetting SWRC				
	θ_s: m³/m³	θ_r: m³/m³	a: m⁻¹	n	m	θ_s: m³/m³	θ_r: m³/m³	a: m⁻¹	n	m
B	0.30	0.10	2.80	1.38	0.28	0.21	0.10	2.20	1.35	0.26
S60	0.29	0.10	4.00	1.50	0.33	0.24	0.10	0.60	1.50	0.33
S120	0.31	0.10	2.10	1.42	0.29	0.26	0.10	1.10	1.35	0.26
S180	0.30	0.10	1.80	1.45	0.31	0.28	0.10	1.80	1.27	0.21

that of the bare soil, and the difference depends on the plant density. For any given suction, the rooted soils in tests S120 and S180 had noticeably greater water retention ability than the bare soil. The presence of roots doubled the AEV of the soil. Interestingly, the rooted soil in test S60 showed reduced, rather than enhanced, water retention ability, as compared with the bare soil (Figure 2.28). The AEV decreased from 2 to 1 kPa, which is against the trend observed in the previous two cases (where planting densities were lower) and is inconsistent with the conceptual model proposed by Scanlan and Hinz (2010). Clearly, more research is needed to investigate how the root growth and decay affect the soil water retention behaviour.

Plant density also had noticeable effects on wetting SWRCs (i.e., those SWRCs obtained during rainfall), as shown in Figure 2.28b. The three vegetated samples had similar adsorption rates (i.e., an increase in VWC for a given decrease in suction), but these rates were lower than that of the bare soil. Because of the hydraulic hysteresis, neither the wetting curve of the bare soil nor that of the rooted soil followed the corresponding drying curve. The hysteresis loop of the rooted soils in test S60 was much larger than those of the other two rooted soil samples and the bare soil. This might be another indication that macropores had formed due to root decay, as the presence of macro-pores would lead to a more open soil structure.

Field monitoring was carried out by Yan and Zhang (2015) to monitor the hydrological responses of a sandy soil with and without *S. heptaphylla* growing in Hong Kong. Pictures of some of the tree seedlings tested at the site at the RCs of 85% and 95% are given in Figure 2.29. When the RC was 85%, the root system was dominated by lateral

(a)   (b)

Figure 2.29 Excavated root systems of *S. heptaphylla* grown in sandy soil at an RC of (a) 85% and (b) 95%. (From Yan, W.M. and Zhang, G.H., *Can. Geotech. J.*, 52, 1–14, 2015.)

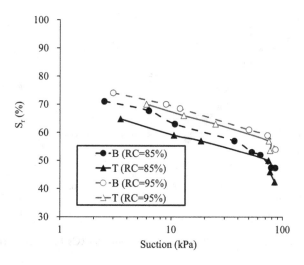

Figure 2.30 Comparisons of SWRCs of bare soil and tree-vegetated soil at different RC levels. (From Yan, W.M. and Zhang, G.H., *Can. Geotech. J.*, 52, 1–14, 2015.)

roots growing mainly in the top 100 mm of soil. The main roots had a diameter of about 18 mm, whilst the secondary roots had a diameter of 5–9 mm. The root systems were much smaller in the soil with an RC of 95%. Comparisons depicted in Figure 2.30 show that regardless of the soil density considered, the SWRC of the vegetated soils always falls below that of the bare soil for any given suction. Interestingly, the findings are contradictory to the test results on grassed soil reported by Rahardjo et al. (2014) (Figure 2.26), but they are consistent with the data presented by Jotisankasa and Sirirattanachat (2017) (see Figure 2.27a). Moreover, Yan and Zhang (2015) showed that tree roots did not affect the desorption rate.

Leung et al. (2015a) conducted a laboratory test by using an experimental setup similar to that shown in Figure 2.7. Although the same CDG and *S. heptaphylla* were used for testing, the RC of the soil considered in Leung et al. (2015a) was 80% (instead of 95%, as described in the tests reported in Section 2.2). Both the bare soil and the vegetated soil were subjected to evaporation and *ET* in the plant room at the HKUST (Figure 2.1), and the changes in soil moisture content and MS were monitored continuously. Figure 2.31a compares the drying SWRCs of the bare and vegetated soils. The AEV of the bare soil was about 1 kPa. It is evident that the AEV of the vegetated soil (3–4 kPa) was higher than that of the bare soil. The observed increase in AEV indicates an increase in water retention capacity, owing to the presence of roots in the soil pore space. This is in line with the experimental observation made by Romero et al. (1999) and Ng and Pang (2000a, 2000b), who also showed that a decrease in the void ratio of the bare soil (i.e., denser soil) would lead to a higher AEV. On the other hand, the bare and vegetated soils had similar desorption rates.

Along the wetting path (Figure 2.31b), neither the increase in VWC in the bare soil nor that in the vegetated soil followed the corresponding drying path, resulting in some hydraulic hysteresis. The wetting SWRCs of the vegetated soil exhibited fluctuations of ±5% owing to the variability in tree root biomass. Based on the repeated tests, it can be seen that the absorption rate (i.e., the increase in VWC due to a decrease in suction) of the vegetated soil was similar to that of the bare soil.

In situ double-ring infiltration tests were performed by Ng et al. (2016d) on a flat compacted CDG (silty sand) with and without *S. heptaphylla* planted in it. A constant water

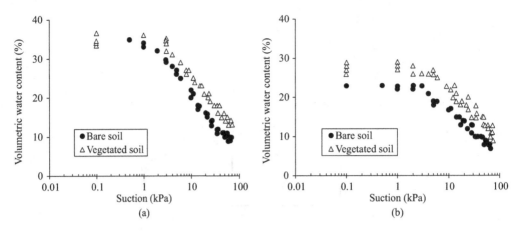

Figure 2.31 Comparison of (a) drying SWRCs and (b) wetting SWRCs between bare and tree-vegetated soil samples.

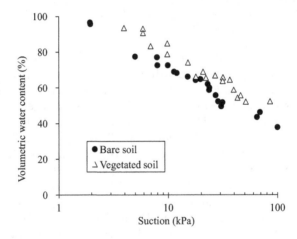

Figure 2.32 Comparisons of SWRCs of bare soil and tree-vegetated soil.

head was applied on the ground surface for infiltration, during which the soil moisture content and MS were continuously monitored to construct SWRCs. Figure 2.32 compares the SWRCs of the bare and vegetated soils. The observed trends for both soils with and without vegetation are fairly similar to those presented in Leung et al. (2015a) – AEV increased, whereas the desorption rate remained unchanged.

2.4.2 Water permeability function

Relatively little test data are available on the effects of plant roots on water permeability function ($k[\psi]$, where ψ is MS) in unsaturated soils. The literature mainly contains the data concerning the effects of roots on the infiltration rate. In general, the existing test data do not seem to reach the consistent conclusion. It was found that infiltration rates in the vegetated soil could be higher (van Noordwijk et al., 1991; Mitchell et al., 1995; Leung et al., 2017b) or lower (Gish and Jury, 1983; Huat et al., 2006; Ng et al., 2014a) than those in the bare soil. Infiltration rates in the vegetated soil were lower when roots were actively growing, but they were higher when mature roots were decaying. However, higher infiltration

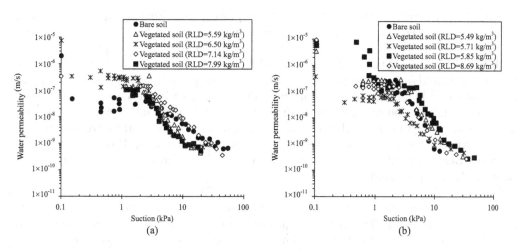

Figure 2.33 Measured effects of plant roots on $k(\psi)$ for (a) silt and (b) clayey sand. (From Jotisankasa, A. and Sirirattanachat, T., *Can. Geotech. J.*, 54, 1612–1622, 2017.)

rates found in natural soil were affected by preferential flow along surface cracks because of excessive soil shrinkage in the field (Simon and Collison, 2002; Ghestem et al., 2011; Leung and Ng, 2013a, 2013b), regardless of plant age.

Jotisankasa and Sirirattanachat (2017) adopted the instantaneous profile method in laboratory studies to measure the effects of roots on $k(\psi)$ for two different types of soil with different root contents (Figure 2.33). Root-induced changes in $k(\psi)$ were prominent mainly for MS below 10 kPa. The test data also suggest that plant roots do not necessarily reduce water permeability (Figure 2.33a). Interestingly, one of the rooted soil samples with RLD of 5.85 kg/m³ in Figure 2.33b exhibited dual-permeability behaviour. When MS was below 1 kPa, the water permeability was up to two orders of magnitude higher than that of the bare soil. Indeed, whether the presence of plant roots would increase or decrease water permeability remains unclear. Some field and laboratory studies (Gish and Jury, 1983; Gabr et al., 1995; Ng et al., 2014a; Rahardjo et al., 2014) showed a reduction in k_s, whereas others observed an increase in k_s, owing to decayed roots (Li and Ghodrati, 1994; Vergani and Graf, 2015) or competition for soil resources such as water (Ng et al., 2016e; Ni et al., 2017, 2018a). Recent modelling work performed by Shao et al. (2017) showed that a dual-permeability model could better capture the hydrological responses of vegetated soils than the conventional single-permeability model.

2.5 CHAPTER SUMMARY

This section shows how plant transpiration and root-induced changes in soil hydraulic properties affect MS and the water infiltration rate in unsaturated soils. The amount of suction induced by grass transpiration depends on soil density. Although looser vegetated soil at an RC of 70% resulted in 57% greater root depth than that at an RC of 95%, the depth of influence of rainfall on suction in the looser vegetated soil was located much deeper in the ground. On the contrary, the 95% compacted vegetated soil had the shallowest roots, leading to the shallowest influence zone of suction. Vegetated soil with an RC of 95% had the greatest ability to retain suction during and after the applied rainfall (it retained 60% of the initial suction after rainfall). This vegetated soil had the lowest infiltration rate.

Plant density plays an important role in the magnitude and distribution of transpiration-induced suction. Reducing the tree spacing from 180 to 60 mm induced greater competition among trees for water, as indicated by a 364% increase in peak suction upon transpiration. Such interaction among trees led to (i) a 19%–35% reduction in LAI, (ii) a 17%–36% decrease in root length and (iii) an obvious decay of roots. During plant transpiration, a higher plant density induced higher suction and yielded a deeper zone of influence of suction. During rainfall, the infiltration rate for soil, with trees planted at a spacing of 60 mm, was up to 247% higher than that for soil with wider tree spacings, where mainly fresh roots were found. Although most of the suction within the root zone (i.e., the top 100 mm) was lost due to an increased infiltration at the 60 mm spacing, suction at greater depths below the root zone was largely preserved.

Plants growing in a nutrient-deficit silty sand (CDG) can survive, but their growth in terms of LAI was inhibited under elevated CO_2 concentration at 1000 ppm, owing to the scarcity of soil nitrogen. This has consequently led to the reduction of plant transpiration and the associated induced soil suction. This is because plants transpire less water when CO_2 concentration is elevated. When sufficient nutrient (N-P-K in a portion of 30%:10%:10% by volume) was supplied, the plant growth of LAI was improved under the high CO_2 concentration of 1000 ppm. Owing to the better growth of leaf, higher suction was induced.

Plant traits, including LAI, RAI, SLA, RLD and RSR showed positive and significant correlations with transpiration-induced suction. These traits may therefore be used as criteria for screening plants for use in soil hydrological reinforcement. No correlation was found between biomass and plant-induced suction, indicating that the latter was instead influenced by other physiological factors such as leaf conductance. In particular, the effect of biomass allocation was highlighted by the positive correlation between the RSR and hydrological reinforcement.

Limited data available in the literature suggest that, in general, the presence of plant roots may modify the water retention ability of the vegetated soil, including the porosity, and hence saturated VWC. However, the effects on AEV and the rate of change of desorption are less clear. Even less data are available for comparing the water permeability functions of soil with and without vegetation, especially when soil is unsaturated. To date, there are only a few such datasets, and even then, it is not easy to isolate and identify root effects (if any) on water permeability function, because of either the lack of experimental accuracy or field uncertainties.

Chapter 3

Mechanical effects of plant root reinforcement

3.1 INTRODUCTION

Roots have been adopted for soil mechanical reinforcement for centuries (Smith and Snow, 2008). Plant roots provide tensile strength to soil, and thus, understanding the biomechanical properties of roots such as tensile strength and Young's modulus is crucial for estimating the extent to which plant roots can mechanically reinforce soil (Wu et al., 1979; Mickovski et al., 2007). The recent state of the art in slope stability improvement can be found in many books (e.g., Coppin and Richards, 1990; Barker, 1995; Gray and Sotir, 1996) and four conference proceedings for the *International Conference of Soil, Bio- and Eco-Engineering: The Use of Vegetation to Improve Slope Stability* (Stokes et al., 2004, 2007; Hubble et al., 2017). Most of these studies focused on the root biomechanical behaviour of plant species native to America and Europe, and in general, they adopted the negative power (or power decay) law as an empirical expression to explain the relationship between root tensile strength (or sometimes root Young's modulus) and root diameter. This chapter provides an update on the root biomechanical properties of some of the plant species native to Asia and also evaluates the applicability of the negative power law to a wide range of root diameters. The possibility of using microorganisms (fungi) to enhance root biomechanical properties is also explored.

3.2 REVISITING THE POWER DECAY LAW (VERBATIM EXTRACT FROM BOLDRIN ET AL., 2017A)

3.2.1 The state of the art

Many studies on root mechanical reinforcement have assessed the relationship between root diameter and root tensile strength (Mattia et al., 2005; Bischetti et al., 2009; Mickovski et al., 2009; Preti and Giadrossich, 2009; Ghestem et al., 2014). The power decay law below has been commonly used to describe the variation in root tensile strength (T_r) with root diameter (d_r) for several plant species:

$$T_r = k_1 d_r^{-k_2} \tag{3.1}$$

$$\log(T_r) = \log(k_1) - k_2 \log(d_r) \tag{3.2}$$

where k_1 and k_2 are positive empirical fitting coefficients to be determined from the tests and they are species dependent. Mao et al. (2012) listed k_1 and k_2 for 81 grass, forb, shrub and tree species reported in the literature. Eq. (3.1) has been used as a model to predict the mechanical reinforcement that can be provided by roots through the so-called root cohesion (Mao et al., 2012), via different existing models, for examples, fibre breakage models (Wu et al., 1979) and fibre bundle models (Pollen and Simon, 2005). However, it should be noted that the power decay law fitting can mainly explain a small fraction of the variability in the tensile strength-diameter relationship (Mattia et al., 2005; Ghestem et al., 2014; Vergani et al., 2014). Biomechanical properties of root change over time as a function of root chemical composition (i.e., cellulose and lignin contents) (Genet et al., 2005; Zhang et al., 2014; Saifuddin et al., 2015), root type (Loades et al., 2013), root age (Dumlao et al., 2015; Loades et al., 2015), root decay (Watson et al., 1999), root disease (Preti, 2013), root moisture content (Yang et al., 2016; Boldrin et al., 2018b) and soil mechanical stress (Chiatante et al., 2003; Loades et al., 2013). Although the tensile strength-diameter relationship has been generally considered to follow a power decay law, the physical basis for this relationship is still unclear nor its optimal use for different species.

3.2.2 Root tensile behaviour of species native to temperate Europe

Ten woody species were selected for root tensile tests to examine the power decay law. These included *Buxus sempervirens* L., *Corylus avellana* L., *Crataegus monogyna* Jacq., *Cytisus scoparius* (L.) Link, *Euonymus europaeus* L., *Ilex aquifolium* L., *Ligustrum vulgare* L., *Prunus spinosa* L., *Salix viminalis* L. and *Ulex europaeus* L. Their family, common name, height range, age and acronym are summarised in Table 3.1. These shrub and tree species have been suggested as suitable plants for soil eco- and bio-engineering applications (Coppin and Richards, 1990; Marriott et al., 2001; Norris et al., 2008; Beikircher et al., 2010) and are suited to a wet maritime Nord European climate. Plants 30–80 cm tall and older than a year were selected for testing.

Following the root sampling and preparation methods to those described in Boldrin et al. (2017a), the root length percentage per diameter class recorded in five replications per species is given in Figure 3.1. *S. viminalis* had the largest percentage of very fine roots (<0.1 mm), exceeding 40%. The other species had a much lower percentage of 20% or less. In most of the tested species, 40%–90% of the total root length measured applied to the root diameter classes between 0.1 and 0.5 mm. In all tested species, roots with diameters larger than 1 mm constituted less than 10% of the total root length.

Table 3.1 The 10 species selected for testing in this study. Their family, common name, height, age and acronym used throughout this study are reported

Species	Family	Common name	Height (cm)	Age (year)
Buxus sempervirens L. (Bs)	Boxaceae	European box	30–40	3
Corylus avellana L. (Ca)	Betulaceae	Hazel	60–80	2
Crataegus monogyna Jacq. (Cm)	Rosaceae	Hawthorn	60–80	2
Cytisus scoparius (L.) *Link* (Cs)	Fabaceae	Scotch broom	40–60	2
Euonymus europaeus L. (Ee)	Celastraceae	Spindle	60–80	2
Ilex aquifolium L. (Ia)	Aquifoliaceae	Holly	40–60	2
Ligustrum vulgare L. (Lv)	Oleaceae	Privet	60–80	2
Prunus spinosa L. (Ps)	Rosaceae	Blackthorn	60–80	1
Salix viminalis L. (Sv)*	Salicaceae	Willow	60–80	1
Ulex europaeus L. (Ue)	Fabaceae	Gorse	40–60	2

Source: Boldrin, D. et al., *Plant Soil*, 2017a.

*indicates propagation by cutting. All plants were supplied by the British Hardwood Tree Nursery, Gainsborough, UK.

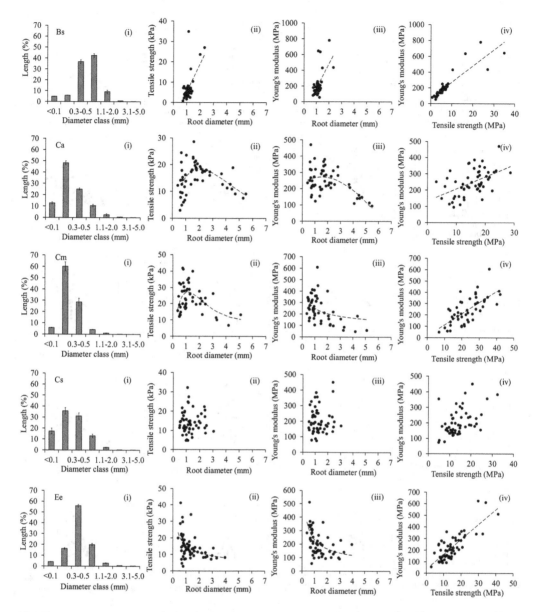

Figure 3.1 Morphological and biomechanical properties of tested species, including percentage of root length in each diameter classes between <0.1 and 5.0 mm (mean ± standard error of mean; n = 5); and root tensile strength plotted against diameter. Dashed lines represent the best-fit curve. (From Boldrin, D. et al., *Ecol. Eng.*, 109, 196–206, 2017a.)   (*Continued*)

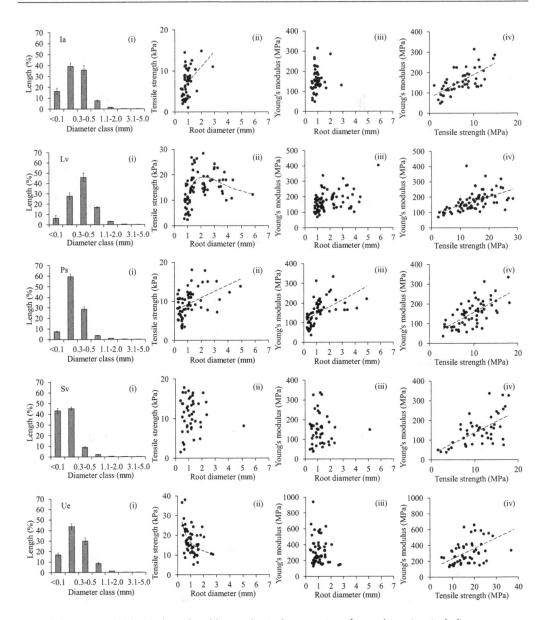

Figure 3.1 (Continued) Morphological and biomechanical properties of tested species, including percentage of root length in each diameter classes between <0.1 and 5.0 mm (mean ± standard error of mean; n = 5); and root tensile strength plotted against diameter. Dashed lines represent the best-fit curve. (From Boldrin, D. et al., *Ecol. Eng.*, 109, 196–206, 2017a.)

Uniaxial root tensile tests were subsequently performed to determine the root tensile strength of each species. Replicates of each root sample were subjected to monotonic uniaxial tension, but at a lower extension rate of 2 mm/min (Boldrin et al., 2017a). The peak tensile force of each root section was identified from a tensile stress-strain curve and the peak tensile strength (T_r) was determined by:

$$T_r = \frac{F_{max}}{A_r} = \frac{4F_{max}}{\pi d_r^2} \tag{3.3}$$

where d_r (mm) is the root diameter at the rupture location (Nilaweera and Nutalaya, 1999; Pollen and Simon, 2005; Tosi, 2007); F_{max} (kN) is the maximum tensile force; and A_r (m²) is the root cross-sectional area at the rupture location. Statistical analysis was performed. Significant differences were assessed with one-way analysis of covariance (ANCOVA), followed by Tukey test. ANCOVA determines whether there are any statistical differences between the means of three or more independent groups. Results were considered statistically significant when the coefficient of significance (p-value) was less than or equal to 0.05.

Figure 3.2 shows that the average tensile strength of species varying from 7.1 ± 0.9 MPa (*B. sempervirens*) to 23.2 ± 1.2 MPa (*C. monogyna*). Moreover, *C. monogyna* had the highest root tensile strength per individual root section (41.8 MPa), which was recorded in a root segment with a diameter of 0.75 mm. The measured root tensile strength-diameter relationships of the 10 species exhibited three different trends, which may be described by three different types of fitting equations (Figure 3.1; Table 3.2). A power decay trend was observed only in *E. europaeus* and *U. europaeus*, while *B. sempervirens*, *I. aquifolium* and *P. spinosa* showed an increase in tensile strength with diameter. Moreover, *C. avellana*, *C. monogyna* and *L. vulgare* seemed to show an initial increase in strength with diameter, but beyond the peak strength, a significant strength reduction with diameter followed. This bimodal trend may be described by a critical exponential equation (Table 3.2). Although the value of R^2 of all three models indicated that only 17%–36% of variation was typically accounted for, the strength-diameter relationship did not always follow a negative power law in several species. *C. scoparius* and *S. viminalis* did not show any significant relationship between root tensile strength and root diameter.

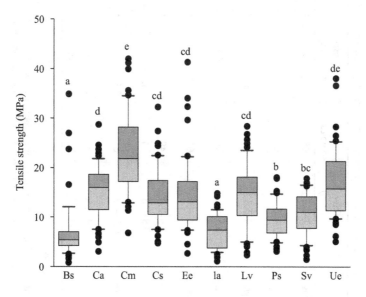

Figure 3.2 Boxplots of root tensile strength for each tested species. The letters at the top indicate significant differences among species, as tested using one-way ANOVA followed by Tukey's test (data are log transformed in statistical analysis). The bottom and top of a box represent the 25th and 75th percentiles, while the line within the box marks the median. Whiskers (error bars) above and below the box indicate the 90th and 10th percentiles. Black circles are outlying points. (From Boldrin, D. et al., *Ecol. Eng.*, 109, 196–206, 2017a.)

Table 3.2 Summary of the data obtained on the root tensile strength and Young's modulus (mean ± standard error of mean) for each tested species

Species	Diameter range (mm)	n samples	Average tensile strength (MPa)	Fitting equation—tensile strength [p-value;Adj. R²]	Average Young's modulus (MPa)	Fitting equation—Young's modulus [p-value; Adj. R²]
Buxus sempervirens L. (Bs)	0.7–2.3	46	7.1 ± 0.9	$f = -8.42 + 13.72*x$ [<0.001;0.35]	211.8 ± 21.7	$f = -144.40 + 315.36*x$ [<0.001;0.35]
Corylus avellana L. (Ca)	0.4–5.6	54	15.3 ± 0.7	$f = -1.9 + (2.9 + 24.9*x)*(0.624^x)$ [<0.001;0.29]	246.7 ± 10.9	$f = -3112 + (3345 + 375*x)*(0.909^x)$ [<0.001;0.28]
Crataegus monogyna Jacq. (Cm)	0.4–5.1	49	23.2 ± 1.2	$f = 8.95 + (-12.7 + 64.2*x)*(0.349^x)$ [<0.001;0.24]	242.3 ± 16.9	$f = 248.63*x^{-0.30}$ [0.003; 0.16]
Cytisus scoparius (L.) Link (Cs)	0.5–3.5	54	14.2 ± 0.8	n.s.	203.1 ± 11.3	n.s.
Euonymus europaeus L. (Ee)	0.4–4.0	61	14.5 ± 0.9	$f = 15.19*x^{-0.46}$ [<0.001;0.18]	221.4 ± 14.8	$f = 232.11*x^{-0.49}$ [<0.001;0.18]
Ilex aquifolium L. (Ia)	0.5–2.9	49	7.2 ± 0.5	$f = 3.66 + 3.66*x$ [0.002;0.17]	157.3 ± 8.3	n.s
Ligustrum vulgare L. (Lv)	0.8–5.8	74	14.7 ± 0.7	$f = 10.41 + (-38.8 + 40.6*x)*(0.451^x)$ [<0.001;0.36]	173.0 ± 7.3	n.s.
Prunus spinosa L. (Ps)	0.2–4.9	61	9.5 ± 0.5	$f = 7.64 + 1.64*x$ [<0.001;0.19]	150.9 ± 8.2	$f = 109.75 + 35.31*x$ [<0.001;0.28]
Salix viminalis L. (Sv)	0.4–5.5	45	10.9 ± 0.6	n.s.	148.9 ± 11.5	n.s.
Ulex europaeus L. (Ue)	0.4–2.7	53	17.2 ± 0.9	$f = 16.61*x^{-0.46}$ [<0.001;0.25]	323.2 ± 22.9	n.s.

Source: Boldrin, D. et al., *Plant Soil*, 2017a.

Note: Best-fit equation, p-values and R^2 are given for the relation between strength and diameter and between Young's modulus and diameter. n.s. indicates the lack of a significant relation. In each fitted equation, x denotes root diameter, while y denotes either tensile strength or Youngs's modulus.

3.2.3 Inter-species variability

The range of root tensile strength found in this study were generally smaller than those reported in the literature. For *S. viminalis*, a maximum tensile strength of 18 MPa was found, which was about eight times lower than that (150 MPa) reported in Mickovski et al. (2009). The differences may be partly explained by the plasticity of root biomechanical properties in response to the growth environmental conditions, such as soil moisture and density (Loades et al., 2013). Indeed, plant adaptive changes (plasticity) enable roots to adjust to spatial and temporal heterogeneity, thus minimizing abiotic and biotic stresses (Stokes et al., 2009). Moreover, the transplanting and the consequent root turn-over (Watson, 1987) might have increased the percentage of younger roots, which are generally weaker than mature roots (Dumlao et al., 2015; Loades et al., 2015).

3.2.4 Strength-diameter relationships

The results highlight that the relationships between root tensile strength and root diameter do not necessarily follow the commonly-quoted power decay law (Figure 3.1). The power decay law applies only to *E. europaeus* and *U. europaeus*. On the other hand, *C. avellana*, *C. monogyna* and *L. vulgare* showed a consistent rapid initial increase in strength with diameter but for thicker roots a significant weakening with increasing diameter followed. This may be explained by the differences between root primary and secondary structures. In the primary structure, the cortex usually occupies the largest volume of roots and consists mainly of highly vacuolated parenchyma cells with diffuse intercellular spaces (Gregory, 2008). In general, the cortex – a parenchymal tissue – is characterized by thin cell walls lacking in cellulose and lignin, the two main structural components contributing to tissue strength (Niklas, 1992; Genet et al., 2005; Zhang et al., 2014). For the tested woody species, thin roots (<1 mm in diameter) can be reasonably assumed to be young and less developed (i.e., primary and early-stage secondary structures). These thin roots are hence expected to have low tensile strength. In contrast, as the proportion of secondary xylem (thick lignified cell walls) increases and the cortex is lost, the biomechanical performance of plant tissues would improve. The development of the secondary structure of root tissues may explain the peak strength for both *C. avellana* and *L. vulgare* for the diameter range of 1.5–2.5 mm.

In general, the root tensile strength at small root diameters (<1 mm) has high variability, as shown in Figure 3.1 and reported in the literature (Mickovski et al., 2009; Ghestem et al., 2014; Zhang et al., 2014). This may be partially explained by the transition from the late stage of the primary structure to the early stage of the secondary structure. During this transition, the cortex is isolated from the rest of the root. Hence the cortex dies and it is sloughed off as part of the normal ageing process (Gregory, 2008). Therefore, primary- and secondary-structure roots may co-exist in the same diameter class, due to the two contrasting processes: secondary xylem development (i.e., the diameter increases) and cortex loss (i.e., the diameter decreases). Despite having the same diameter, such roots have different tissue compositions and hence varying biomechanical properties. In this case, the power decay law for the root tensile strength-diameter relation may not be applicable as it could substantially overestimate root mechanical reinforcement, especially for very fine to fine roots which are dominant in all of the tested species in this study.

The post-peak decrease in tensile strength with increasing root diameter is in agreement with most of the literature (Genet et al., 2005; Bischetti et al., 2009; Mickovski et al., 2009; Loades et al., 2013; Ghestem et al., 2014). Zhang et al. (2014) explained that decreases in tensile strength resulted from a decrease in the lignin-to-cellulose ratio with increasing diameter. The observed strength decrease may also be explained by an increase in weak points, from which fractures propagate, as root diameter increases. Moreover, perennial roots may experience environmental stresses such as waterlogging or droughts in certain years leading to localized effects on tissues and biomechanical properties (Cutler et al., 2009; Loades et al., 2013). In particular, droughts and waterlogging can affect the wood structure. There is a strong correlation between soil water deficit and wood density, which can be attributed to a decrease in xylem vessel enlargement and an associated increase in the proportion of cell walls in wood tissue (Bouriaud et al., 2005). On the contrary, wet growing conditions result in xylem vessels of larger diameters and a decrease in wood strength (Arnold and Mauseth, 1999; Alam et al., 2015). Since the post-peak relationship between root strength and diameter generally accords with the literature, the negative power law may be adopted with some confidence for roots with a diameter larger than 2 mm (i.e., the common cut-off between fine and thin roots; Stokes et al., 2009). Applying this cut-off in the power decay law may prevent the over-prediction of root strength. Świtała et al. (2018) reported that grass fine roots could also increase the shear strength of soil. More research is needed to improve the understanding of fine roots in terms of their tensile strength and influence on grassed soil such as erosion resistance.

3.3 ROOT TENSILE BEHAVIOUR (VERBATIM EXTRACT FROM LEUNG ET AL., 2015)

3.3.1 Four plant species native to southern China

Figure 3.3 shows four native species, *Rhodomyrtus tomentosa* (Rh), *Melastoma sanguineum* (Me), *Schefflera heptaphylla* (Sc) and *Reevesia thyrsoidea* (Re), commonly found in Hong Kong (Hau and So, 2003). They have a very high survival rate, which is vital for improving and maintaining slope stability (Hau and Corlett, 2003), in addition to high ornamental and ecological value with attractive flowers and/or fruits. Rh and Me are shrubs while Sc and Re are small trees (Hu and Wu, 2008). Shrubs are woody, perennial plants that are smaller in size than trees (Harris et al., 2004; AFCD, 2010). They have shorter woody stems branching readily to form a clump of foliage that lacks a definite crown shape (Koptur et al., 1988; Harris et al., 2004). They usually reach a height of 1~2.5 m in the early succession stage and grow further to 2.5~4 m if conditions are favourable (Thrower, 1984; Kochummen et al., 1990). Trees are also woody, perennial and long-living plants. But unlike shrubs, they usually have a single, erect main trunk (AFCD, 2010; Lilly and Currid, 2010). Their branches grow at least 1 m above the ground (Koptur et al., 1988) and form a characteristic crown of foliage once they reach maturity (Thrower, 1984; AFCD, 2010; BSI, 2010). Small trees are those that are less than 9 m tall (Cheung, 2007). The native species studied in this chapter appear in the list of plant species recommended for ecological slope greening in Hong Kong (Hau and So, 2003, 2005a, 2005b; GEO, 2011). The two native shrub species are common pioneer species on Hong Kong's shrubland. They have a short lifespan of about 20–50 years (GEO, 2011).

Rh is an evergreen shrub species of the family Myrtaceae that can grow on natural and man-made slopes (Hau et al., 2005). It has a mature height of 1–2 m (Hu and Wu, 2008). Its leaves vary from elliptic to obovate in shape with a broadly cuneate base, a round to obtuse apex and obvious reticulate veins (Hu and Wu, 2008). Flowers appear in late spring to summer and are purplish red in colour (Figure 3.3a). Berries are fleshy and purplish black when mature in late summer to autumn. They attract wildlife, especially birds (Corlett, 1998, 2001).

Me is also an evergreen shrub species of the family Melastomataceae. It has a height of 1.5–3 m (Hu and Wu, 2008). Its leaves are papery and have a lanceolate or elliptic-lanceolate shape. Two pairs of lateral primary veins can be found on each side of the mid-veins (Hu and Wu, 2008). The species grows fleshy capsules with relatively high carbohydrate and water contents, which attract birds (Figure 3.3b; Corlett, 2001). It is also highly tolerant of drought conditions and has a high survival rate as shown by previous experimental results (Hau et al., 2005).

Figure 3.3 Seedlings of four selected species native to southern China: (a) *R. tomentosa*, (b) *M. sanguineum*, (c) *S. heptaphylla* and (d) *R. thyrsoidea*. (From Leung, F. T. Y., The use of native woody plants in slope upgrading in Hong Kong, PhD thesis, The University of Hong Kong, Hong Kong, 2014.)

Sc is an evergreen tree species of the family Araliaceae. It can grow up to 10 m at maturity (but normally 5 m only) (Hu and Wu, 2008). Their leaves are digitately compound, with 6–8 leaflets. Their leaflets are rather leathery, varying from elliptic to ovate-elliptic in shape (Figure 3.3c). The species propagates well in natural and man-made environments. It has good vigour and is highly tolerant of adverse environmental conditions, such as drought (Hau and So, 2003, 2005a). Ecologically, it is a keystone species for forest restoration (Hau and So, 2003) and an important source of nectar for bumblebees and butterflies as well as a source of fruit for birds from late summer onwards and throughout the dry season (Corlett, 1998, 2001).

Re is a semi-deciduous tree species of the family Sterculiaceae. It normally grows up to 5 m. Its leaves vary from lanceolate to ovate-lanceolate in shape and have a leathery texture (Hu and Wu, 2008). It has high ornamental value, thanks to its dense white-petal flowers that blossom in spring (Figure 3.3d). It is therefore a popular species for enriching the floristic composition of woodlands (AFCD, 2012), and is suitable for the ecological greening of slopes (Or et al., 2011). Many butterfly and moth species are popular visitors for the nectar (Corlett, 2001).

3.3.2 Root sampling and measurement of root area ratio (RAR)

The root systems of the four selected species were sampled from hillsides in Tai Tam and Lung Fu Shan Country Parks as well as the Kadoorie Centre of the University of Hong Kong between 2011 and 2013. For all species, plants ranging from 1.0 to 1.5 m tall were sampled. The living root system of each plant was carefully retrieved manually using hand-held tools.

The root systems of three Rh, five Me, six Sc and nine Re specimens were examined to reveal their root distribution and variation in RAR with depth. RAR is defined as (Gray and Leiser, 1982; Gray and Sotir, 1996)

$$RAR = \frac{A_r}{A} = \frac{\sum_{i=1}^{n_r} \pi d_i^2 / 4}{A} \tag{3.4}$$

where d_i is the diameter of the i-th root among a total of n_r identified roots within an area A, which is the fraction of soil cross-sectional area; and A_r is the root cross-sectional area. Any broken/lost root segments during the excavation and retrieval processes would cause the RAR to be underestimated but they give a conservative shear strength of the rooted soil. It is assumed that RAR varies exponentially with depth (z) as follows:

$$RAR = h_1 z^{-h_2} \tag{3.5}$$

$$\log(RAR) = \log(h_1) - h_2 \log(z) \tag{3.6}$$

where h_1 and h_2 positive empirical fitting coefficients to be determined from test data. These two coefficients are species-dependent.

Figure 3.4 shows the distribution of root diameter classes with depth for the four species. Generally, the roots of shrubs (Rh and Me) are finer than those of trees. Besides, the root system of tree species (Sc) extends deeper into the ground. Re comprises fewer roots and exhibits mainly a tap root system. Figure 3.5 shows the variation in RAR with depth for diameters ranging from 1 to 10 mm. It can be seen that RAR reduced noticeably with depth. For all studied species, RAR became insignificant below a depth of 0.5 m. Furthermore, roots having a diameter of 1–10 mm had an RAR that was only one-third of that when roots of all diameters were considered. Moreover, the root class distributions shown in the figure reveals that the root system of the studied species (except Me) was dominated by roots larger than 2 mm in diameter. Levene's test showed that the variance in RAR of different species was, statistically, not homogeneous. This variability may be due to genetics and/or environmental factors such as soil moisture, nutrient and soil compaction. RARs of the studied species were similar to those found in the Southern French Alps (approximately 0.05%; Burylo et al., 2011) but significantly smaller than those found in Mediterranean species (mostly below 0.5% but can be as high as 2% for grasses and shrubs) reported by De Baets et al. (2008). In their study, the reference area A (Eq. 3.4) for the calculation of RAR is the area under the crown of a plant.

Table 3.3 summarises the ANCOVA of RAR with depth as a covariate. All RAR data satisfied the normality requirement but only some fulfilled the homogeneity of variance requirement. It can be seen that the RAR between the two shrubs was significantly different ($F_{1,38} = 5.76$, p-value = 0.021). Moreover, the RAR between the two trees was similar ($F_{1,135} = 3.92$, p-value = 0.07). Also, the RAR between shrubs and trees (e.g. Rh and Re) was significantly different (Rh: $F_{4,24} = 3.80$, p-value = 0.016; Re: $F_{8,87} = 4.05$, p-value < 0.001).

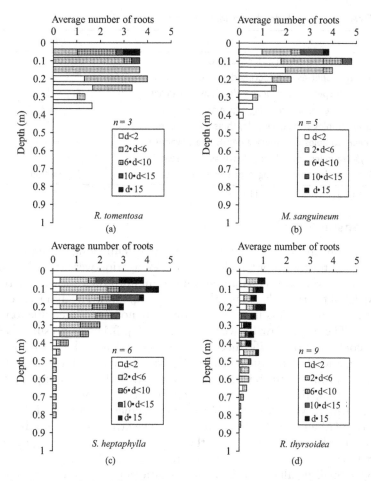

Figure 3.4 Root distribution with depth for (a) *R. tomentosa* (*n* = 3), (b) *M. sanguineum* (*n* = 5), (c) *S. heptaphylla* (*n* = 6) and (d) *R. thyrsoidea* (*n* = 9). (From Leung, F. T. Y. et al., *Catena*, 125, 102–110, 2015.)

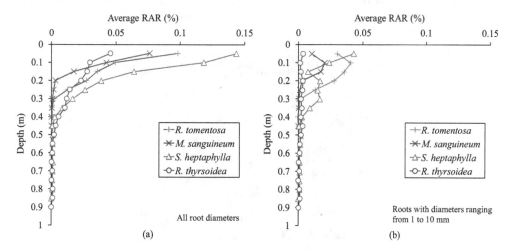

Figure 3.5 Distribution of root area ratio (RAR) with depth for (a) all roots; and (b) roots with diameters ranging from 1 to 10 mm. (From Leung, F. T. Y. et al., *Catena*, 125, 102–110, 2015.)

Table 3.3 ANCOVA of RAR, with depth considered as a covariate

Comparisons	F	*p-value*
Among all species considering depth as a covariate	***	
Between shrub and tree species considering depth as a covariate	***	
Between the two shrub species considering depth as a covariate	$F_{1,38} = 5.76$	0.021
Between the two tree species considering depth as a covariate	$F_{1,149} = 3.92$	0.07
Among *R. tomentosa* samples considering depth as a covariate	***	
Among *M. sanguineum* samples considering depth as a covariate	$F_{4,24} = 3.80$	0.016
Among *S. heptaphylla* samples considering depth as a covariate	***	
Among *R. thyrsoidea* samples considering depth as a covariate	$F_{8,87} = 4.05$	<0.001

Source: Leung, F.T.Y. et al., *Catena*, 125, 102–110, 2015.

3.3.3 Root tensile force

Fresh roots of the four species were cut into segments 150 mm long for tensile strength tests (Figure 3.6). During testing, root tensile force and elongation were recorded continuously. The elongation speed was kept constant at 8 mm/min until the root was ruptured. Tensile strength (T_r, MPa) of the root fibre was calculated by Eq. (3.3). The diameter of tested roots ranged from 1 to 10 mm. Roots finer than 1 mm were not considered due to uncertainties in identification. The power law between T_r and d_r (Eq. 3.1) was presumed. Adjusted R^2 values and the coefficient of significance (*p*-value) were calculated to indicate the goodness of fit, considering a level of significance of 0.05. To compare T_r among species and within the same species (taking diameter as a covariate), ANCOVA was applied to the linear regression of log (T_r) against log (d_r), with a significance level of 0.05.

A total of 121 Rh, 31 Me, 86 Sc and 112 Re root segments were tested. Figure 3.7 shows the variation in log (T_r) with log (d_r) for each species. The fitting coefficients and the statistical significance of the relationships based on the roots with diameters ranging from 1 to 10 mm are summarised in Table 3.4. A linear decrease in log (T_r) with log (d_r) was observed, as indicated by a small *p*-value. The k_2-values of different species were close to each other (varying from 0.816 to 0.830) except those of Me. This means that for all of the selected species, there exists a power relationship between T_r and d_r, for root diameters ranging from 1 to 10 mm. Since root samples with diameter smaller than 1 mm were not available (Leung et al., 2015), whether the root tensile strength in small diameter range follows the negative

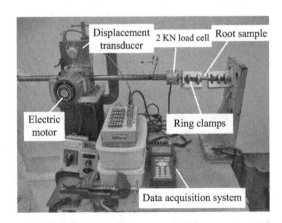

Figure 3.6 Test setup for measuring root tensile force (From Leung, F. T. Y. et al., *Catena*, 125, 102–110, 2015.)

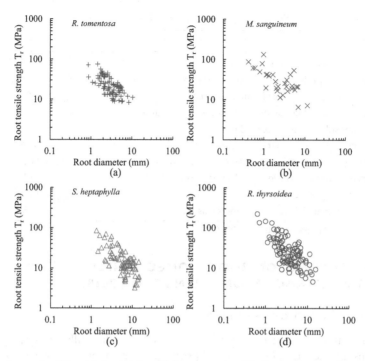

Figure 3.7 Variation in log(T$_r$) with log(d$_r$) for (a) *R. tomentosa*, (b) *M. sanguineum*, (c) *S. heptaphylla* and (d) *R. thyrsoidea*. (From Leung, F. T. Y. et al., *Catena*, 125, 102–110, 2015.)

Table 3.4 Summary of regression analyses of root tensile strength with root diameter

Species	No. of samples	log(k$_1$)	k$_2$	Adjusted R^2	p-value
R. tomentosa (Rh)	87	1.695	0.816	0.448	<0.001
M. sanguineum (Me)	19	1.561	0.513	0.433	0.001
S. heptaphylla (Sc)	55	1.844	0.826	0.529	<0.001
R. thyrsoidea (Re)	85	1.800	0.830	0.481	<0.001

Source: Leung, F. T. Y. et al., *Catena*, 125, 102–110, 2015.

power law (Eq. 3.1) cannot be commented at this stage. The fitted log (k_1) value of each species varied within a narrow range from 1.561 to 1.844, which correspond to k_1 of 36.4 and 69.8, respectively. The coefficient k_2 varied narrowly from 0.816 to 0.830 for Rh, Sc and Re, while Me had a smaller k_2 of 0.513.

Table 3.5 summarises the ANCOVA of root tensile strength with diameter as a covariate. The requirements for normality and homogeneity of variance were always satisfied. The analysis revealed no significant difference in tensile strength between the two shrub species or between the two tree species. However, the root tensile strengths of the studied shrubs and trees were statistically different. Roots of the tree species (Sc and Re) had higher resistance to tension than those of the shrub species (Rh and Me). On the contrary, some studies (Gray and Barker, 2004; De Baets et al., 2008) reported comparable tensile strengths between shrubs and trees, with others observing a higher tensile strength of French shrub species than trees (Burylo et al., 2011). The inter-species variation in tensile strength is high as it is expected to depend largely on the local environment. Local research on the biomechanics of native species is thus crucial.

Table 3.5 ANCOVA of root tensile strength, with root diameter considered as a covariate

Comparisons	F	p-value
Among all species considering diameter as a covariate	$F_{3,241} = 7.66$	<0.001
Between shrub and tree species considering diameter as a covariate	$F_{1,243} = 21.34$	<0.001
Between the two shrub species considering diameter as a covariate	$F_{1,103} = 0.03$	0.866
Between the two tree species considering diameter as a covariate	$F_{1,137} = 1.76$	0.187
Among all species considering diameter <2 mm	$F_{3,39} = 1.48$	0.234
Among all species considering diameter >2 mm	$F_{3,198} = 5.74$	0.001
Within *R. tomentosa* samples considering diameter as a covariate	$F_{1,84} = 3.65$	0.059
Within *M. sanguineum* samples considering diameter as a covariate	$F_{1,16} = 2.61$	0.126
Within *S. heptaphylla* samples considering diameter as a covariate	$F_{1,52} = 0.16$	0.694
Within *R. thyrsoidea* samples considering diameter as a covariate	$F_{1,82} = 3.96$	0.050

Source: Leung, F.T.Y. et al., *Catena*, 125, 102–110, 2015.

3.4 EFFECTS OF FUNGI ON ROOT BIOMECHANICS (VERBATIM EXTRACT FROM CHEN ET AL., 2018)

3.4.1 Actions of fungi on cellulose

Conventional maintenance of geotechnical structures such as landfills and slopes does not take living organisms into account. However, soil properties can change over time due to ecological processes (DeJong et al., 2013, 2015) and this should not be ignored. Plants, bacteria, fungi, and burrowing animals/insects (such as ants and earthworms) can have considerable influence on the properties of the plant–soil system, such as root tensile strength (a function of root cellulose content) (Wu et al., 1979; Genet et al., 2005), the amount of organic matter, void ratio, soil aggregate stability, permeability, and soil water retention characteristics. These properties subsequently would affect induced suction and stability of the soil structure (DeJong et al., 2013). Thus, biological factors should be considered in the design of geotechnical works.

Arbuscular mycorrhizal fungi (AMF) have been one of the most important microorganisms associated with plants since 400 million years ago (Smith and Read, 2008). These symbiotic fungi have profound influences on soil properties and plant growth. For example, AMF generate significant amounts of organic matter, namely glomalin, which creates cohesion among soil particles and attracts other living soil organisms (Bedini et al., 2009). Increasing plant biomass upon AMF symbiosis is commonly reported (Klironomos, 2003; Smith and Read, 2008; Bonfante and Genre, 2010). More importantly, it has been shown that the expression of glycosyltransferase (which is responsible for cellulose biosynthesis) could be promoted through inoculation with AMF (Fiorilli et al., 2009), although this effect could be plant and AMF species dependant (Klironomos, 2003).

3.4.2 Effects of the AMF colonisation rate on plant biomass

To study the soil-plant-fungus interaction, vetiver grass *Chrysopogon zizanioides* was selected as the host as it possesses a significant amount of root biomass and forms a symbiotic association with AMF (Wong et al., 2007). The inocula of three species of AMF, *Rhizophagus intraradices* (Ri), *Funneliformis mosseae* (Fm) and *Glomus aggregatum* (Ga), obtained from Beijing Academy of Agriculture and Forestry Sciences were used as the symbionts of Vetiver grass. Completely decomposed granite (CDG), which was compacted at relative compaction of 95%, was used as the growth substrate for the grass. Twenty grams of the inoculum of each arbuscular mycorrhizal fungus was added uniformly to the soil surface. Three batches of vetiver grass of approximately the same size (about 200 mm tall) were transplanted to a pot each with the roots in direct contact with the inoculum. Sterilized inoculum (autoclaved at 121°C for 2 h) was added to another five pots as a control (non-mycorrhizal, NM). The same volume of the half-strength Hoagland nutrient solution (Hoagland and Arnon, 1950) with limited phosphorus (one-fifth of concentration) was added to each pot every two weeks. Subsamples of the root were collected and stained, and the colonisation rate of the AMF was assessed using the slide length method (Giovannetti and Mosse, 1980).

The AMF colonisation rates were less than 5% on average for all treatments. These low levels of colonisation, relative to the 50%–80% found in Wong et al. (2007), did not prevent the AMF from contributing to enhancing root wet biomass (Figure 3.8a), while the shoot wet biomass only increased upon inoculation with Ga. Ri and Fm had moderate effects on the root and shoot dry biomass, whereas Ga significantly promoted both the root and shoot dry biomass (Figure 3.8b and c). Increases in plant biomass are commonly observed upon AMF treatment (Wong et al., 2007; Smith and Read, 2008). In this study, fungal structures including arbuscules, vesicles, and hyphae were observed in the cortex of the roots (Figure 3.9).

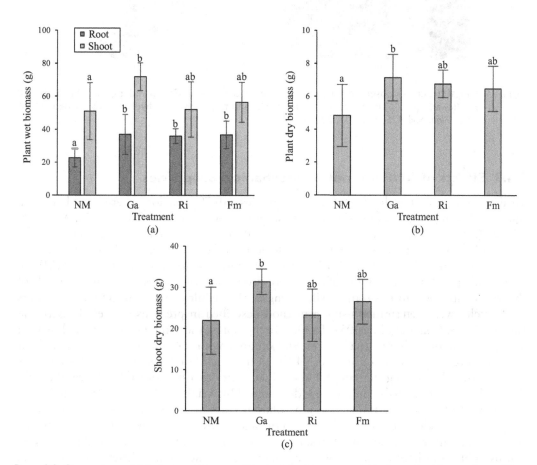

Figure 3.8 Comparison of (a) plant wet biomass, (b) plant dry biomass and (c) shoot dry biomass from different treatments. The letters at the top indicate significant differences among treatments. Duncan's multiple range test at a probability level of 5% is used for post-hoc comparison to separate the differences. Data are presented as mean ± standard deviation (n = 5). (From Chen, X.W. et al., *Plant Soil*, 425, 309–319, 2018.)

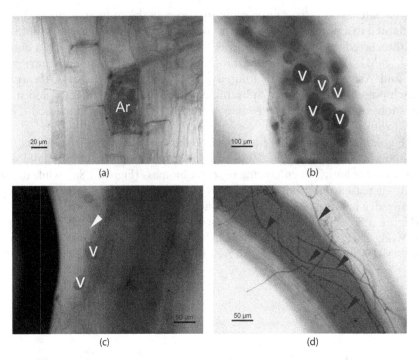

Figure 3.9 Fungal structures observed in roots inoculated with the AMF Ga ((a) and (b)), Ri ((c) and (d)) and Fm ((e) and (f)). "a" is arbuscule, "v" is vesicle, and the arrows are hyphae. (From Chen, X. W. et al., *Plant Soil*, 425, 309–319, 2018.)

3.4.3 Effects of AMF on root biomechanical properties

To determine root tensile strength, root samples 60–70 mm long were tested using a universal testing machine operated at a constant rate of 4 mm/min (Tosi, 2007). A total of 170, 242, 257 and 225 pieces of NM, Ga, Ri and Fm roots were tested, respectively. The force and diameter of the roots (and the stele) at breakage were recorded to determine the peak tensile strength using Eq. (3.3). To assist the interpretation of the effects of AMF on root tensile properties, a number of other root properties were measured. Since the stele of the root contributes to the root tensile strength, the cellulose and hemicellulose contents in the steles were determined using a method described in previous studies (Leavitt and Danzer, 1993; Genet et al., 2005). To remove the lipids, the samples were treated using a Soxhlet extractor with 99% toluene and 96% ethanol at a ratio of two to one (2–1; v/v) for 24 h, followed by ethanol for another 24 h. On the other hand, the hydrosoluble contents were removed by heating the samples in distilled water to 100°C for 6 h. Hence, the total mass of lignin and that of cellulose were determined by weighing them. To eliminate lignin, the samples were further immersed in a solution containing 700 mL of distilled water, 7.0 g of sodium chlorite and 1.0 mL of acetic acid and heated to 70°C for 12 h. Finally, the samples with the sachets were dried and weighed. The cellulose, hemicellulose, lipid, hydrosoluble, and lignin contents were calculated from the relative difference before and after extraction.

All statistical analyses were performed using version 22 of the Statistical Package for the Social Sciences (SPSS) software. Normality of data was checked using the Shapiro-Wilk test. Data were log-transformed to meet the normality assumptions where necessary. Levene's test

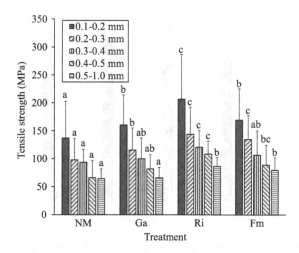

Figure 3.10 Comparison of root tensile strengths derived from different mycorrhizal treatments and different classes of stele diameters. The letters at the top indicate significant differences between treatments for each class of stele diameter. Duncan's multiple range test at a probability level of 5% is used for post-hoc comparison to separate differences. Data are presented as mean ± standard deviation. (From Chen, X. W. et al., *Plant Soil*, 425, 309–319, 2018.)

was used to assess the homogeneity of data. To separate the difference among treatments, one-way analysis of variance (ANOVA) was conducted followed by Duncan's multiple range test at a significance level of 5%.

For the stele diameter ranges of 0.1–0.2 and 0.2–0.3 mm (Figure 3.10), all AM treatments significantly increased (*p*-value < 0.05) the tensile strength. For the diameter range of 0.3–0.4 mm, only the Ri treatment increased the tensile strength. For the diameter ranges of 0.4–0.5 and 0.5–1.0 mm, both the Ri and Fm treatments enhanced the tensile strength, but not the Ga treatment. These results indicated that the AM treatments could generally enhance the stele tensile strength, especially the tensile strength of fine stele (0.1–0.3 mm). This is because AMF can easily colonise fine roots, rather than coarse roots (Wu et al., 2016), and exert local effects in the colonised cortical cells (Fiorilli et al., 2009). The fungal structures were only found in fine roots in this study. Higher percentages of cellulose and hemicellulose contents were observed in the samples treated with AMF (letters above error bars, Figure 3.11). The proportions of lipid and hydrosoluble contents were reduced, and this was compensated only by cellulose and hemicellulose, but not lignin.

AMF symbiosis may enhance cellulose and hemicellulose biosynthesis without affecting lignin production. The increase in root dry mass was moderate (Figure 3.8b), while the increases in tensile strength (Figure 3.10) and cellulose and hemicellulose contents (Figure 3.11) were more significant than those in the control. A decrease in the proportion of a certain type of content does not necessarily imply a decrease in the actual amount. Since the total mass increased upon AMF inoculation, even though the proportions of lipid and hydrosoluble contents decreased, their actual amounts could still increase.

Plants respond differently to colonisation by various AMF. In a previous study, nearly half of the 64 plant species inoculated with *Glomus etunicatum* showed increments in biomass, while the other half showed decrements (Klironomos, 2003). It is not clear whether AMF symbiosis would universally enhance the proportions of cellulose and hemicellulose contents in different plant species.

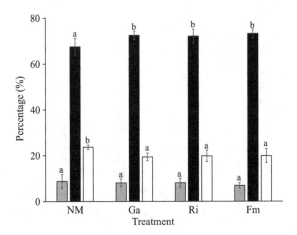

Figure 3.11 Comparison of the percentages of lignin (gray bar), cellulose + hemicellulose (black bar), and lipid + hydrosoluble contents (white bar) in the steles of roots derived from different mycorrhizal inoculations. Duncan's multiple range test at a probability level of 5% is used for post-hoc comparison to separate differences. Data are presented as mean ± standard deviation. ($n = 5$). (From Chen, X. W. et al., Plant Soil, 425, 309–319, 2018.)

3.4.4 Potential mechanisms

Root biomass can be promoted through inoculation with AMF (Smith and Read, 2008), which would increase the total cellulose and hemicellulose contents. The proportion of each type of content may remain unchanged. However, the results showed not only increments in the actual amount, but also percentage increments. The percentages of cellulose and hemicellulose contents were positively correlated with root tensile strength (Genet et al., 2005). Thus, the density and bonding of cellulose microfibrils (composed of glucan chains) and xylan chains (hemicellulose) in each unit of the cross-sectional area may be essential parameters. The increase in the cell wall thickness during AMF colonisation could be caused by cell wall loosening required for the fungal intracellular colonisation and/or synthesis of new structural polymers (Balestrini and Bonfante, 2005). The wall extensibility may be related to cellulose-cellulose contacts, which are considered as biomechanical 'hot spots' (Cosgrove, 2014). The root strength is derived from the organisation of cellulose microfibrils which is related to the degree of polymerization and microfibril crystallinity (Saxena and Brown, 2005; Joshi and Mansfield, 2007).

Upon AMF colonisation, more cellulose synthase (CesA) complexes, which synthesise cellulose microfibrils, may be assembled in the Golgi apparatus (Wightman and Turner, 2010) and this will lead to a greater number (or higher density) of complexes transferred to and activated in the plasma membrane. The following factors are involved in this process: (1) enhanced CesA expression (a CesA catalyses the reaction that assembles glucose into a glucan chain), (2) enhanced efficiency of the Golgi apparatus in complex production, (3) enhanced trafficking of the complexes from the Golgi apparatus to the plasma membrane (Bashline et al., 2014), (4) activation of more complexes and (5) increased photosynthesis of glucose for glucan chain production. Such a process may increase the number of rosette subunits (formed by six CesA proteins) and hence the number of rosettes (formed by six rosette subunits) located in the plasma membrane. The CesA assembles more glucose into a glucan chain, and the rosettes are responsible for synthesising these glucan chains into cellulose microfibrils via hydrogen bonding (Taylor, 2008).

Hemicellulosic polysaccharides have been grouped into xyloglucans, xylans, mixed-linkage glucans, mannans and glucomannans (Scheller and Ulvskov, 2010; Pauly et al., 2013). Unlike cellulose, hemicelluloses – with a degree of polymerization of only 200 – is synthesised through different routes. For commelinid monocots, such as vetiver grass, the xylan contents (specifically glucuronoarabinoxylan, GAX) are the major hemicelluloses in the primary and secondary cell walls which compose 20%–40% and 40%–50% (w/w) of the tissues, respectively (Scheller and Ulvskov, 2010). Mutants with abnormal xylan biosynthesis have a collapsed xylem phenotype which impairs the resistance to the high negative pressure generated by transpiration pull (Pauly et al., 2013). This indicates that xylan, in addition to glucan in cellulose, contributes to the tensile strength as well.

With the involvement of AMF, an alternative mode of cellulose and hemicellulose biosynthesis has yet to be proposed. The test data based on the use of bacteria and plant mutants (mostly *Arabidopsis*) did not indicate the involvement of such a ubiquitous symbiosis between plant and AMF in the natural environment. *Arabidopsis* is considered as a non-mycorrhizal plant (Veiga et al., 2013), and therefore it is probably not suitable for studying cellulose and hemicellulose biosynthesis upon mycorrhizal symbiosis.

The structure of the plant cell wall changes upon the colonisation of AMF. When the cortex is colonised by AMF, the hydroxyproline-rich glycoprotein (HRGP) and cellulose are expressed at two sites: the wall and the interface area created by invagination of the host membrane around the developing fungus (Bonfante et al., 1990; Balestrini et al., 1994). In contrast, in uninfected roots of the same age, HRGP and cellulose are only present in the inner part of the wall. A specific antibody against β-1,3-glucans demonstrates that these glucans are not laid down at the interface between the plant and the fungus, while they appear to be a skeletal component of the fungal wall, together with chitin (Balestrini et al., 1994). Cell wall material is laid down between the host plasma membrane and the fungal cell surface (Balestrini and Bonfante, 2014). However, only the cell wall in the cortex (where AMF colonisation occurs) is investigated in the literature. Information about the cell wall in the stele, in which the cellulose content increases, is lacking. The increase in cellulose content appears to be a whole plant phenomenon, since cellulose contents increase in the stele where AMF colonisation does not occur.

The three AMF tested exerted significant effects (p-value < 0.05) on root wet biomass and the proportions of cellulose and hemicellulose contents in the root stele of vetiver grass. This finding was confirmed by the enhancement of root tensile strength. Ri was the most effective at increasing root tensile strength, followed by Fm and Ga. The three AMF tended to increase the tensile strength of roots with smaller stele diameters (0.1–0.3 mm). They played an important role in the formation of the plant cell wall. Future studies should take AMF into account to obtain a more complete picture of cellulose and hemicellulose biosynthesis.

3.5 CHAPTER SUMMARY

This chapter reveals the root biomechanics of four woody species native to Hong Kong and 10 other woody species widespread in the wet maritime climate Nord European region. The relationship between root tensile strength and diameter does not necessarily follow the commonly assumed power decay law, especially for small to very fine roots (i.e., root diameters less than 1 mm). A comparison of the root biomechanics of species native to regions exhibiting different climates suggests that the relationship between root tensile strength and root diameter is species dependent. The four species native to Hong Kong, Rh, Me, Sc and Re, follow the power decay law for relatively large root diameters, ranging from 1 to 10 mm. However, out of the 10 European species investigated in this chapter, the law is applicable to only two species, *E. europaeus* and *U. europaeus*, for root diameters ranging from 0.03 to 5 mm. On the contrary, *B. sempervirens*, *I. aquifolium* and *P. spinosa* showed a slight increase in tensile strength with diameter. Interestingly, *C. avellana*, *C. monogyna* and *L. vulgare* all showed an initial increase in root tensile strength with diameter, reaching peak strengths of 1.0–2.5 mm diameter. Beyond the peak strength, a reduction in strength was observed. This bimodal trend might be partially explained by the differences in the development stage of root primary and secondary structures.

Indiscriminate use of the negative power law to describe the root tensile strength–diameter relation could overestimate root mechanical reinforcement, especially for roots with a very small diameter, for which the model would likely predict a high tensile strength. Caution should be exercised when this root biomechanical model is assumed for different taxa or for the same species growing under different environmental conditions or during the challenging initial establishment period.

This chapter also shows that AMF could increase the percentage of cellulose and hemicellulose contents in plant roots (p-value < 0.05). Cellulose and hemicellulose are the main structural components contributing to tissue strength such as root tensile strength. The increase in cellulose and hemicellulose was confirmed by the enhanced tensile strength (which is positively correlated with the proportions of cellulose and hemicellulose contents) of the stele. Mycorrhizae tended to improve the tensile strength of steles with smaller diameters (0.1–0.3 mm) to a greater extent. The AMF could also reduce the lipid and hydrosoluble contents (p-value < 0.05) but not the lignin content.

Chapter 4

Field studies of plant transpiration effects on ground water flow and slope deformation

4.1 INTRODUCTION

Plant transpiration has been shown to have significant effects on soil hydrological changes in terms of soil moisture and matric suction in flat vegetated soil under controlled-atmosphere conditions (Chapter 2). This chapter aims to extend the discussion of the effects of plant community (e.g., mixed tree–grass planting) and vegetation management (e.g., tree spacing) on plant growth, soil infiltrability and transpiration-induced changes in slope hydrology under natural field conditions. Assuming that the behaviour of unsaturated soil is governed by two independent stress-state variables (net stress and matric suction; Fredlund and Rahardjo, 1993; Ng and Menzies, 2007), effects of transpiration-induced changes in suction, soil stress and slope movement are investigated by interpreting three well-documented field case studies. The three field cases include (1) a compacted sandy ground at the HKUST Eco-Park, (2) a cut slope in expansive clay in Zaoyang, China, and (3) a natural saprolitic hillslope in Hong Kong.

4.2 CASE STUDY 1: COMPACTED SANDY GROUND AT HKUST ECO-PARK (VERBATIM EXTRACT FROM LEUNG ET AL., 2015B; NI ET AL., 2017)

A test site, called HKUST Eco-Park (Figure 4.1), was developed by the authors for investigating the plant effects on the infiltration rate and the effects of mixed grass–tree species planting on matric suction change in unsaturated soil. The site had a compacted, 2 m high by 10 m wide embankment and two slope angles of 22° and 33°. The soil used to construct the embankment was completely decomposed granite (CDG; well-graded sand with silt, ASTM, 2011). The embankment was compacted to the dry density of 1.8 g/cm^3, about 95% of the maximum dry density (i.e., relative compaction of 95%). Some basic properties of the soil are described in Table 2.1 and also by Garg et al. (2015a), Leung et al. (2015b) and Ni et al. (2017).

4.2.1 Plant effects on the infiltration rate

On the top flat surface of the embankment, a series of double-ring infiltration tests were carried out in compacted soil with and without vegetation. Two plant species were considered, a grass (*Cynodon dactylon*, also known as Bermuda grass) and a tree species (*Schefflera heptaphylla*, also known as Ivy tree), for which the hydrological effects of transpiration-induced

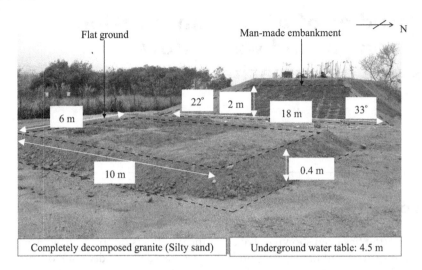

Figure 4.1 Overview of the HKUST Eco-Park.

matric suction and root biomechanical properties (for *S. heptaphylla* only) have been discussed in Chapters 2 and 3, respectively. Figures 4.2a and b show an overview of the grass specimens (one year after germination) and a tree individual after one year of growth at the site, respectively. The tree root system had a tap root with a diameter of 6 mm and a cluster of fine roots with diameters ranging from 1 to 2 mm.

Figures 4.3a and b show the schematic and overview of a typical setup of a doublering infiltration test, respectively, and instrumentation for the test plot vegetated with selected tree species. The setup consisted of an inner ring (0.3 m in diameter), an outer ring (0.6 m in diameter) and a calibrated Mariotte's bottle for maintaining constant water head inside both rings. Each infiltration test followed the test setup described in ASTM (2009), with the only difference being that the grass and tree specimens were transplanted inside both rings. An array of Jet-Fill tensiometers (JFTs) was installed for measuring pore-water pressure (PWP) or suction change during testing. To consider any effects of *in situ* soil heterogeneity and plant variability on water infiltration, three replicated tests (Inf1, Inf2 and Inf3) were conducted. According to test procedures outlined in ASTM (2009), water was filled inside both the inner and outer rings and the water head in both rings was maintained by the Mariotte's bottle for 2 h for allowing water infiltration.

Figure 4.4 compares the infiltrated water volume and water infiltration rate with time between bare, grass-covered and tree-covered soil during the 2 h of ponding from test Inf1. In the bare soil, the volume of water infiltrated increased but at a decreasing rate. After ponding for 1 h (equivalent to a rainfall with return period of 100 years in Hong Kong; DSD, 2013), the volume of water infiltrated increased linearly with time, indicating that the steady-state condition had likely been reached. A similar trend of variation was observed for both the grass- and tree-covered soils. The volumes of water infiltrated in these two types of vegetated soils were similar. Although the levels of initial suction in the bare, grass-covered and tree-covered soils were comparable (less than 25% difference, Leung et al., 2015b), up to 50% less water infiltrated the vegetated soils than the bare soil at the steady state. This indicates that the presence of plant roots can significantly reduce the volume of water infiltration. As shown in the figure, the infiltration rate in both bare and vegetated soils decreased exponentially at a reducing rate.

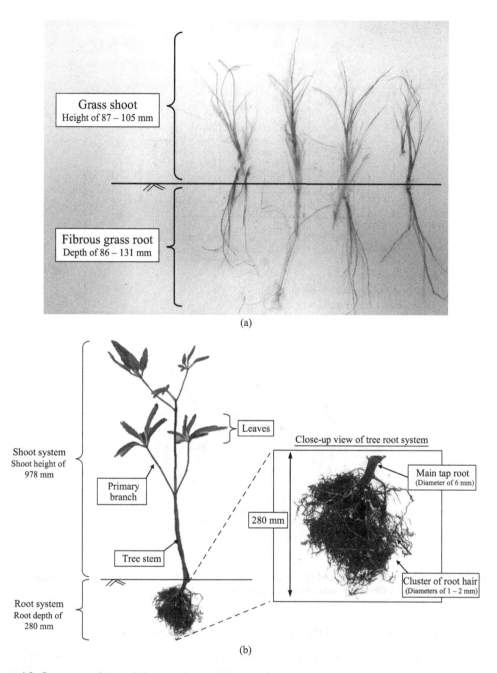

Figure 4.2 Overview of typical shoot and root systems of the selected (a) grass species, *Cynodon dactylon*, and (b) tree species, *Schefflera heptaphylla* tested in series Infl.

At the steady state, the infiltration rate approached the saturated water permeability (k_s) of the CDG. The infiltration rate in the grass-covered soil was always lower than that in the bare soil. The difference was the largest (45%) at the beginning of the test, but this difference was reduced as ponding progressed with time. The infiltration rates in the grass-covered and tree-covered soils did not seem to differ from each other throughout the ponding event.

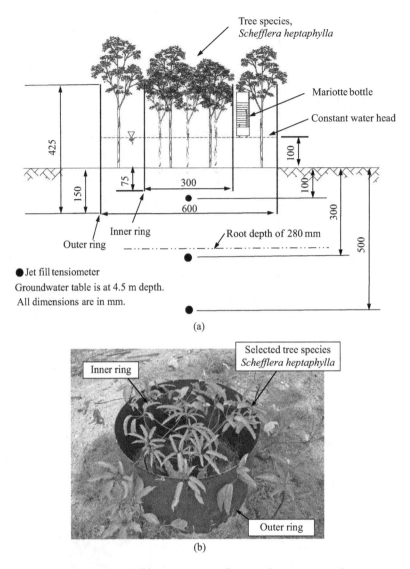

(a)

(b)

Figure 4.3 (a) A schematic diagram and (b) an overview of a typical test setup and instrumentation for the double-ring infiltration tests in tree-covered soil.

4.2.2 Effects of plant variability on the infiltration rate

Figure 4.5 compares the measured upper and lower bounds of infiltration rates in the bare and vegetated soils tested in the three repeated tests Inf1, Inf2 and Inf3. The infiltration rates measured in the bare soil in tests Inf1-B, Inf2-B and Inf3-B varied to some extent. The maximum difference between the lower bound and upper bound of infiltration rates was relatively large during the first 20 min of the test, before gradually reducing as ponding progressed with time. Given that the soil density and the initial suction were similar in the replicated tests, the observed variation was attributed to the *in situ* soil heterogeneity (i.e., spatial variability) at the three locations tested at the site.

Both the upper and lower bounds of the infiltration rates in the grass-covered soil were lower than those in the bare soil. The observed variations in the infiltration rate in the

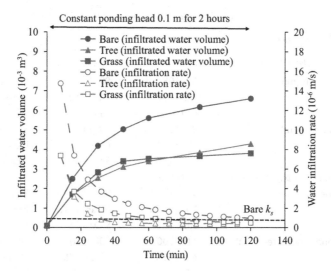

Figure 4.4 Measured variations of cumulative volume of water infiltrated and infiltration rate with time for bare, grass-covered and tree-covered soil in series Infl.

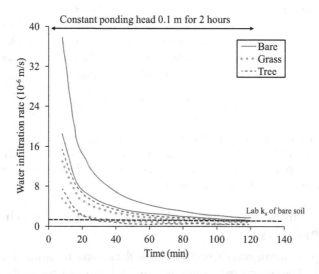

Figure 4.5 Comparisons of upper and lower bounds of infiltration rates for each type of soil tested in the three repeated series Infl, Inf2 and Inf3.

grass-covered soil were attributed to not only *in situ* soil heterogeneity but also some variabilities among grass roots. The upper and lower bounds of infiltration rates correspond to grass having the shortest (Inf3-G; 75 mm) and longest (Inf1-G; 131 mm) root depths, respectively (Table 4.1). When the root depth of grass increased by 75%, the infiltration rate was reduced by half during the first 20 min of ponding. This suggests that grass with longer roots might occupy more soil space, and hence block more channels for water flow during infiltration. This finding supports the assumption to be made for the soil water retention curve model of root-permeated soils in Chapter 5. The infiltration rates in the tree-covered soil were also lower than those in the bare soil. Although similar infiltration rates were

Table 4.1 Summary of grass and tree traits tested in Inf1, Inf2 and Inf3

Plant characteristics	Inf1	Inf2	Inf3
Cynodon dactylon (grass)	Inf1-G	Inf2-G	Inf3-G
Shoot length (mm)	87–105	89–108	86–106
	Mean: 96	Mean: 100	Mean: 94
	S.D.: ±5	S.D.: ±6	S.D.: ±7
Root depth (mm)	86–131	79–113	75–108
	Mean: 112	Mean: 92	Mean: 85
	S.D.: ±10	S.D.: ±10	S.D.: ±8
Schefflera heptaphylla (tree)	Inf1-T	Inf2-T	Inf3-T
Shoot height (mm)	950–1000	900–1100	800–1050
	Mean: 978	Mean: 1013	Mean: 955
	S.D.: ±13	S.D.: ±55	S.D.: ±68
Plant canopy diameter (mm)	110–200	100–220	90–180
	Mean: 153	Mean: 145	Mean: 127
	S.D.: ±18	S.D.: ±34	S.D.: ±30
Leaf area index	0.8–1.3	1.0–2.1	0.6–1.7
	Mean: 1.1	Mean: 1.8	Mean: 1.2
	S.D.: ±0.16	S.D.: ±0.3	S.D.: ±0.3
Root depth (mm)	262–305	254–320	211–260
	Mean: 280	Mean: 291	Mean: 232
	S.D.: ±22	S.D.: ±20	S.D.: ±12
Peak root area index	0.64–0.94	0.72–1.28	0.59–0.75
	Mean: 0.80	Mean: 0.97	Mean: 0.68
	S.D.: ±0.11	S.D.: ±0.15	S.D.: ±0.05

Abbreviation: S.D., standard deviation.

identified in the grass-covered and tree-covered soils in the test Inf1 (Figure 4.4), *in situ* soil heterogeneity and natural variability of trees could result in higher infiltration rates in the tree-covered soil (see upper bound from test Inf3-T) than in the grass-covered soil (see upper bound from test Inf3-G), albeit by not more than 10%.

4.2.3 Effects of mixed tree–grass planting on plant growth and soil hydrology

A field monitoring programme was carried out on the flat compacted ground at the HKUST Eco-Park (see Figure 4.1). The compacted ground was divided into four test plots, each with an area of 1.5 × 1.5 m in plan (Figure 4.6). Three test plots were vegetated with the selected grass and tree species, while the remaining plot was left bare for reference. The three plots were covered with turf made up of *C. dactylon*. The initial length of the grass shoots was 25 ± 12 mm. Then, tree individuals with a similar height of 600 ± 38 mm were transplanted to the three vegetated plots with three different tree spacings (120, 180 and 240 mm). In total, 240 tree individuals were transplanted to the three vegetated plots. Before the field monitoring, the vegetation in all three plots was allowed to grow for 4 months and was irrigated every 3 days in order to maintain an average matric suction similar to that at the field capacity of the CDG (i.e., 23–25 kPa).

The field monitoring was divided into two stages. The first stage was aimed at the mixed-species grounds. Prior to testing, the soil was wetted by water ponding at the ground surface until all tensiometers recorded zero readings. All plots were then left exposed to the atmosphere for monitoring any suction change for 7 days from 14 to 21 April 2015. At the end

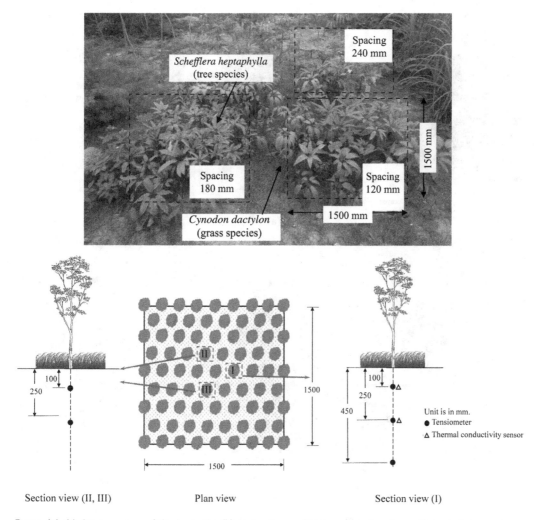

Figure 4.6 (a) An overview of the test site; (b) test setup and instrumentation for each test plot.

of the monitoring, grass shoots in all plots were cut and removed carefully without affecting the shoots of trees and the root systems of both the trees and grasses. After grass removal, the second stage of testing commenced. Similarly, all plots were first wetted with water, followed by another 7 days of monitoring from 22 to 29 April 2015.

4.2.3.1 Observed plant traits

Figure 4.7a shows the effects of tree spacing on the tree root system. As the tree spacing decreased from 240 to 120 mm, the root volume (determined by three-dimensional image analyses) dropped by about 30% (see Table 4.2). This was likely because when the trees were closer together, tree–tree competition intensified, which might have suppressed the growth of tree roots, resulting in a smaller root volume. During the careful excavation of tree roots, decayed roots were identified for the case of a close tree spacing of 120 mm as a result of intense tree-to-tree competition, consistent with the laboratory observations shown in Figure 2.10. On the contrary, relatively few decayed roots were found for the cases of wider spacing. On the other

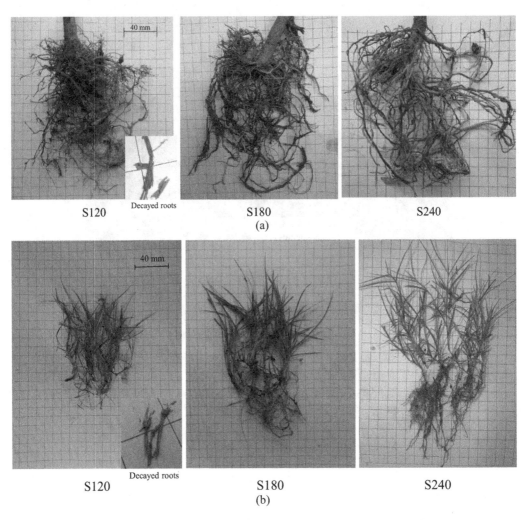

Figure 4.7 Root systems of (a) *Schefflera heptaphylla* and (b) *Cynodon dactylon* at different tree spacings (each grid represents a 10mm*10mm square).

hand, the grass specimens grown under the closest tree spacing of 120 mm had ~48% shorter shoots and ~45% shorter roots than those grown under the widest tree spacing of 240 mm (Figure 4.7b; Table 4.2). Decayed grass roots were also found for the case of a 120 mm tree spacing. The observed reduced grass growth and root decay were likely due to the limited sunlight beneath tree canopies (Scholes and Archer, 1997). Intense grass–tree competition for resources such as water and nutrients, owing to the close tree spacing, might have also reduced the production of the above- and below-ground biomass of grass (Rozados-Lorenzo et al., 2007). This suggests that for the selected grass and tree species at least, the grass–tree competition might outweigh the beneficial effects of trees (i.e., alteration of resource availability and microclimatic conditions; Grouzis and Akpo, 1997) on grass growth. It is clear that research on optimal planting space for different types of species is urgently needed.

Figure 4.8 shows the variations in the leaf area index (LAI) of the grass and tree specimens with time during the four-month growing period under mixed planting condition. The initial mean LAI of the trees and grass was 0.72 ± 0.13 and 0.36 ± 0.12, respectively.

Table 4.2 Summary of measured traits of the tree and grass

	Plant traits	Spacing = 120 mm		Spacing = 180 mm		Spacing = 240 mm	
		Tree	Grass	Tree	Grass	Tree	Grass
Before transplantation	Shoot height (mm)	562–640 Mean: 601 S.D.: ±38	14–32 Mean: 24 S.D.: ±8	568–625 Mean: 596 S.D.: ±26	15–36 Mean: 26 S.D.: ±9	572–632 Mean: 604 S.D.: ±27	12–37 Mean: 24 S.D.: ±12
	Leaf area index	0.58–0.95 Mean: 0.73 S.D.: ±0.13	0.28–0.42 Mean: 0.35 S.D.: ±0.05	0.65–0.81 Mean: 0.75 S.D.: ±0.06	0.25–0.51 Mean: 0.37 S.D.: ±0.12	0.58–0.79 Mean: 0.70 S.D.: 0.08	0.27–0.44 Mean: 0.36 S.D.: 0.07
	Root depth (mm)	100–134 Mean: 116 S.D.: ±18	21–40 Mean: 32 S.D.: ±10	98–140 Mean: 121 S.D.: ±21	20–43 Mean: 29 S.D.: ±11	101–129 Mean: 113 S.D.: ±13	20–45 Mean: 33 S.D.: ±12
Four months after growth	Shoot height (mm)	679–734 Mean: 708 S.D.: ±27	41–68 Mean: 56 S.D.: ±13	675–748 Mean: 715 S.D.: ±35	65–93 Mean: 79 S.D.: ±14	583–742 Mean: 711 S.D.: ±30	90–121 Mean: 108 S.D.: ±16
	Leaf area index	1.20–1.43 Mean: 1.30 S.D.: ±0.11	0.23–0.46 Mean: 0.36 S.D.: ±0.09	1.46–1.64 Mean: 1.54 S.D.: ±0.07	0.58–0.79 Mean: 0.67 S.D.: ±0.10	1.48–1.72 Mean: 1.63 S.D.: ±0.09	0.78–0.99 Mean: 0.89 S.D.: ±0.09
	Root depth (mm)	145–192 Mean: 171 S.D.: ±23	43–61 Mean: 52 S.D.: ±8	168–204 Mean: 186 S.D.: ±16	61–80 Mean: 72 S.D.: ±9	177–216 Mean: 198 S.D.: ±19	81–108 Mean: 96 S.D.: ±14
	Root volume ($\times 10^6$ mm^3)	1.8–2.4 Mean: 2.1 S.D.: ±0.2	N/A	2.1–3.0 Mean: 2.6 S.D.: ±0.4	N/A	2.6–3.3 Mean: 3.0 S.D.: ±0.3	N/A

Abbreviations: S.D., standard deviation; N/A, not applicable.

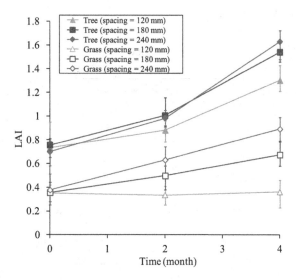

Figure 4.8 Measured changes in LAI of tree and grass specimens during the growing period under mixed planting condition.

When the trees were widely spaced (i.e., 240 mm), the tree LAI doubled during the 4-month growth period. On the contrary, the trees grown closer to each other (i.e., 120 mm apart) had smaller increases in LAI. This led to a final tree LAI of 1.3 ± 0.11, which was 15%–25% lower than that recorded in trees grown further apart. This was likely because of the more significant tree–tree competition for soil moisture and nutrients when trees were planted closer together. Also, the close tree spacing caused tree canopies to overlap, which suppressed the growth of tree leaves underneath. There was no discernible difference in the tree LAI between the spacings of 180 and 240 mm, because the tree canopies did not overlap in these two cases.

Tree spacing showed significant effects on the grass LAI. The grass LAI remained unchanged when the tree spacing was close (i.e., 120 mm). As a result of the overlapping tree canopies, the grass underneath was almost completely shaded from the sun, hindering its growth (Rozados-Lorenzo et al., 2007). Such shading effects reduced with an increase in tree spacing. Therefore, for wider tree spacing, more sunlight would reach the grass, facilitating grass growth (i.e., a higher grass LAI; see Figures 2.12 and 2.13).

Figure 4.9 shows the distributions of root area index (RAI) of the grass and tree specimens with depth under mixed planting conditions. The RAI profiles of trees at any tree spacing exhibited a similar parabolic shape, and all of them consistently peaked at a depth of about 90 mm. Increasing the tree spacing from 120 to 240 mm resulted in an increase of about 15% in root depth (Figure 4.7; Table 4.2). The tree RAI in the top 110 mm of soil was more substantially affected by tree spacing than that below in the three cases. The peak RAI for the closest tree spacing of 120 mm, which was associated with the smallest root volume, was 14%–20% higher than those for the other two wider spacings. The significant tree–tree competition at close tree spacing might have activated the production of abscisic acid for tree root proliferation within the root zone and survival (Munns and Sharp, 1993), thus resulting in the substantial increase in RAI.

The RAI profiles of the grass were markedly different from those of the trees. The grass RAI reduced with depth almost linearly and the peak RAI was always found at the ground surface. The grass RAI was strongly affected by the tree spacing. The peak RAI of the grass increased by 29% as tree spacing increased from 120 to 240 mm, demonstrating evident interaction between the grass and trees. As the trees were planted further apart, the grass–tree competition reduced, resulting in better grass growth and leading to a higher LAI, a higher RAI and a greater root depth.

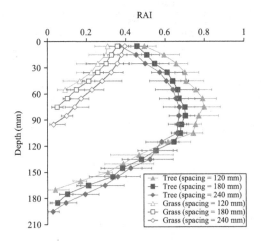

Figure 4.9 Measured profiles of RAI of tree and grass specimens after a four-month growing period under mixed planting condition.

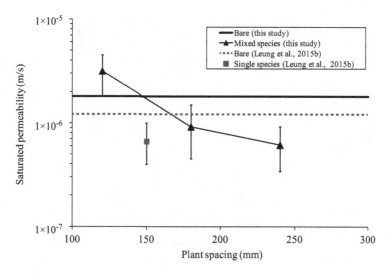

Figure 4.10 Effects of plant spacing on the saturated water permeability of soil.

4.2.3.2 Effects of tree spacing on saturated water permeability (k_s)

Figure 4.10 shows the effects of tree spacing on k_s of the vegetated soils. k_s of the vegetated soils could be higher or lower than that of the bare soil, depending on the tree spacing. k_s clearly reduced as tree spacing increased. At a wider tree spacing (i.e., 180 and 240 mm), the vegetated soils had lower k_s than the bare soil. A plausible explanation for the consistent trend was that the roots had blocked the soil pore space (see Section 2.4 for more details). Another possible reason was that the hydrophobic organic compounds released by roots, such as sugars, amino acids and phospholipids (Bengough, 2012), may have also reduced the k_s observed in the vegetated grounds. This was especially the case for the tree spacings of 180 and 240 mm, where decayed roots were less evident.

In contrast, the measured k_s of the vegetated soil at the spacing of 120 mm was evidently higher than that of the bare soil. The observed decayed roots for this spacing (Figure 4.7) might have created preferential flow channels (Ghestem et al., 2011), causing an increase in the soil water permeability. This effect might have outweighed the effect of the root blockage of soil pore space or the release of hydrophobic organic compounds on k_s.

4.2.3.3 Effects of tree spacing on the transpiration-induced suction response

Figure 4.11a compares the measured variations in suction of the bare and vegetated grounds at a depth of 100 mm with time during the first stage of monitoring (i.e., under mixed grass–tree conditions). Suction induced in any vegetated ground was always higher than that induced in the bare ground owing to transpiration, regardless of the tree spacing. Among the three vegetated grounds, the one with the closest tree spacing (120 mm) showed the highest rate of suction increase and induced the highest suction with a peak of 87 kPa. This might have been because of the markedly higher tree RAI (hence higher root surface area for water uptake) when trees were grown closer together (see Figure 4.9). In this case, grass root water uptake was likely minimal because of the significant suppression of grass growth (see Figures 4.7b, 4.8 and 4.9). A further increase in tree spacing to 180 mm, however, resulted

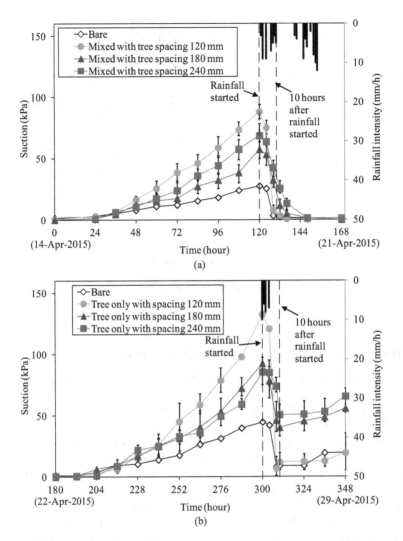

Figure 4.11 Measured variations in matric suction at depth of 100 mm with time (a) with and (b) without grass.

in a peak suction (57 kPa; see Figure 4.11a) that was 16% lower than that in the case of a 240 mm spacing (68 kPa), even though the tree RAI for the 180 mm spacing was 5% higher. Note that an increase in tree spacing will lead to an increase in the grass RAI but a reduction in the tree RAI. This means that as the tree spacing increases, the effects of grass root water uptake become more prominent, affecting suction more significantly.

The suction induced by trees in the absence of grass was generally higher than that in the mixed-species planting case (Figure 4.11b). Plants consume water mainly during metabolic processes, and there may be more water available when inter-species competition is less intense (Qiu et al., 2011; Höltta and Sperry, 2012). This implies that trees would have a higher metabolic rate in the absence of grass. This explains the greater demand for water and hence higher suction in the absence of grass than in the grass–tree mixed condition.

Comparing the suction responses of the three vegetated grounds, the peak induced suction increased with a decrease in tree spacing. Since grass was absent, the major factor contributing to the amount of suction induced would be related to the tree RAI. The higher the tree RAI, the higher the peak suction that would be induced. Such correlation was not found when grass was present (Figure 4.11a). This highlights the significance of grass–tree interaction. Under mixed grass–tree conditions, a wider tree spacing helped reduce grass–tree competition and encourage the growth of grass roots (Figures 4.8 and 4.9), resulting in the combined effects of grass and tree root water uptake on induced suction.

4.2.3.4 Effects of tree spacing on the suction response during rainfall

When grass was present, after a particularly intense episode of rain with a peak intensity of 12 mm/h, no suction was preserved in both the bare and vegetated grounds, regardless of the tree spacing (Figure 4.11a). The closer the trees were, the higher the induced suction before rainfall but the greater rate of reduction in suction. Suction took 16 h to completely vanish in the vegetated ground with the closest spacing of 120 mm, whereas it took 8 h longer to do so in the vegetated grounds with the wider spacings of 180 and 240 mm, because of the increase in k_s (Figure 4.10). During the monitoring period when grass was absent (Figure 4.11b), the rainfall was less intense and lasted for a shorter period, and thus the suction drop was less significant. Certain amounts of suction were preserved in all three vegetated grounds and they were higher than those in the bare ground, regardless of the tree spacing. The closer the tree spacing, the lower the suction preserved.

Figure 4.12 shows the distributions of suction with depth before and after 10 h of rainfall. Under mixed grass–tree conditions (Figure 4.12a), suction in the top 250 mm of all vegetated soils was affected by the rainfall, whereas that at a depth of 450 mm remained unchanged. The presence of vegetation appeared to have negligible effects on the depth of influence of suction, which was consistently shallower than the depth of 450 mm in all three cases. At depths both within and below the root zone of trees, the magnitude of suction preserved was higher for a wider tree spacing. The same trend was found in the absence of grass (see Figure 4.12b). This seems to suggest that in addition to the magnitude of plant-induced suction before rainfall, the effects of tree spacing on k_s also affected the suction responses during rainfall significantly (Figure 4.10). More detailed discussion of how the presence of plant roots would affect the soil water retention curve (Figures 2.26 through 2.32) as well as the saturated and unsaturated permeability (Figure 2.33) can be found in Section 2.4.

Findings in Figure 2.15, which also depict the tree spacing effects on suction but measured under a controlled laboratory environment, suggest that a closer tree spacing helps preserve more suction below the root zone after rainfall. The field study described herein, however, did not provide supportive evidence. In the field study, under mixed grass–tree conditions, the suction preserved both within and below the root zone of trees was the lowest when trees were closely spaced (Figure 4.12a). Instead, a wider tree spacing helped reduce grass–tree competition, encourage grass growth (Figures 4.8 and 4.9), and minimise the number of decayed roots (Figure 4.7), leading to a greater reduction in k_s (Figure 4.10). Clearly, more research is needed to explore and determine optimal planting spacing of vegetation for preserving soil suction and lowering water permeability to enhance the stability of soil slopes.

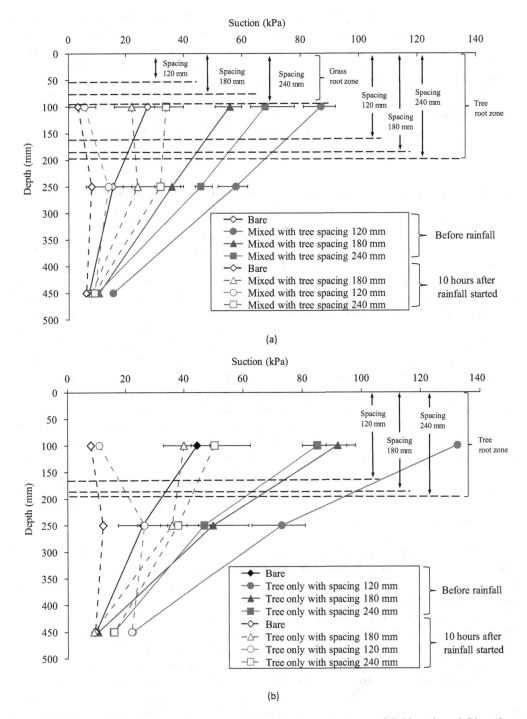

Figure 4.12 Measured distributions of matric suction with depth during rainfall (a) with and (b) without grass.

4.3 CASE STUDY 2: A CUT SLOPE OF EXPANSIVE CLAY SLOPE IN ZAOYANG, CHINA (VERBATIM EXTRACT FROM NG AND ZHAN, 2007)

The test site of this field case study was located in Zaoyang, Hubei, about 230 km northwest of Wuhan and about 70 km south of the intake canal for the South-to-North Water Transfer Project in Nanyang, Henan (Ng et al., 2003). The area was semi-arid with an average annual rainfall of about 800 mm and 70% of the annual rainfall was distributed between May and September. The measured daily temperatures varied between 20°C and 35°C throughout the field study period (i.e., from August to September 2001). The measured daily evaporation potential from a free water surface during the monitoring period ranged from 3 to 10 mm.

Two test areas were selected on a cut slope of an excavation canal (Figure 4.13). The slope angle was 22° and had a height of 11 m. The slope surface was originally covered with grass but no trees were present. The types of grass mainly included weed and couch grass and their heights ranged from 100 to 500 mm. The depths of roots observed on an excavated face ranged from 100 to 300 mm. A bare test site was obtained by removing the top soil to a depth of 100 mm. The soil type found in the cut slope was silty clay with an intermediate plasticity and a medium expansive potential (Zhan, 2003). More detailed properties of the soil profile are summarised in Table 4.3.

Table 4.3 Soil profile and geotechnical properties from the boreholes at mid-slope

Depth (m)	Soil profile	In situ water content (%)	Dry density (g/cm³)	SPT-N[a]	Undrained shear strength (kPa)	K_0[b]
0–1.0	Yellow–brown stiff clay with cracks	14.8–18.8	1.5–1.58			
1.0–1.5	Yellow–brown mottled grey clay	19.5–22.1	1.61	15.0–18.6	83.0	2.8
1.5–3.1	Yellow–brown mixed grey clay	20.0–23.3	1.49–1.62	15.4–17.8	76.8	2.0
3.1–4.1	Yellow–brown clay with many iron concretions	21.4–23.8	1.56	15.0	103.7	1.9
4.1–6.1	Yellow–brown mottled grey clay with small iron concretions	20.5–25.5	1.61–1.62	13.7–18.2	66.2–99.5	1.1–1.6
6.1–7.1	Yellow–brown clay with many iron concretions	21.9–27.2	1.60	19.8–25.1	41.3	0.8
7.1–8.0	Yellow–brown mixed grey clay	21.2–24.6	1.55	19.0–19.4	124.3	1.3
8.0–10.0	Yellow–brown clay with coarse calcareous concretion	22.6–23.4				

[a] SPT-N denotes the number obtained from the standard penetration tests.
[b] K_0 values were measured by dilatometer tests (DMT).

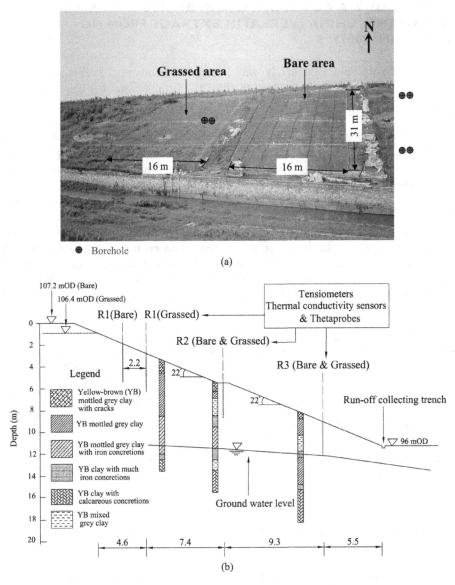

Figure 4.13 (a) Overview of the research slope in Zaoyang, Hubei, China and (b) cross-section of the instrumented slope. (Ng, C. W. W. and Zhan, L. T., *Soils Found.*, 47, 207–217, 2007.)

The upper soil layer with a thickness varying from 1.0 to 1.5 m had many cracks and fissures. Trial pit exploration near the monitoring area (Figure 4.14) showed that in the upper part of the pit, there were four major predominately vertical cracks. The maximum width of the open cracks was about 15 mm. Numerous randomly distributed hair-like fissures were also observed. On the exposed face in the lower part of the trial pit, two major cracks were observed with widths that decreased as depth increased. The maximum crack depth was about 1.2 m from the soil surface. The abundance of cracks and fissures was mainly attributed to the desiccation of the expansive clay during dry seasons.

In both bare and grassed areas, 12 tensiometers and 12 thermal conductivity sensors were installed in three rows namely, R1, R2 and R3, in the upper, middle and lower parts

Figure 4.14 Cracks and fissures in the excavation pit near the monitoring area (the max. depth of cracks ≈ $d_1 + d_2 = 1.2$ m): (a) upper part; (b) lower part. (Ng, C. W. W. and Zhan, L. T., *Soils Found.*, 47, 207–217, 2007.)

of the slope, to monitor changes in PWP and suction (Figure 4.13b). Moisture probes (Thetaprobe, 1999) were also installed in both areas for monitoring changes in the volumetric water content (VWC). Most of the sensors were embedded within a depth of 2 m. In this case study, rainfall was produced artificially using a sprinkler system designed by Ng and Zhan (2007). As shown in Figure 4.15, the artificial rainfall in the bare area lasted for 7 days. The measured daily potential evaporation from a free water surface during the period ranged from 3 to 10 mm. On the other hand, the artificial rainfall applied in the grassed area also lasted for about 7 days. The intensity of artificial rainfall in the grassed area was close to that in the bare area. No natural precipitation was recorded during the artificial rainfall period. The total rainfall depth in the grassed area was about 380 mm.

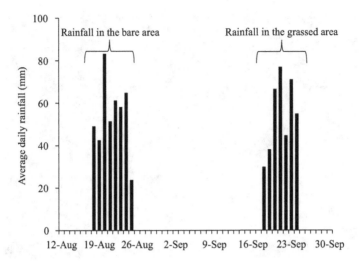

Figure 4.15 Rainfall events simulated in the bare and grassed areas. (Ng, C. W. W. and Zhan, L. T., *Soils Found.*, 47, 207–217, 2007.)

4.3.1 Grass effects on infiltration characteristics

Figure 4.16 compares the measured infiltration rates of the bare and grassed slopes with a similar gradient of 22°. Right after rainfall commenced, the infiltration rate was equal to the rainfall intensity of 2.9 mm/h, since no surface runoff was collected. In the bare area, the infiltration rate started to decrease after 1.5 days of rainfall. On the contrary, in the grassed area, the infiltration rate did not decrease until after 3 days of rainfall. This meant that surface runoff in the grassed area showed a longer delay. During the one-week rainfall event, the infiltration rate in the grassed area was higher than that in the bare area for four possible reasons. First, the grass cover intercepted and absorbed a small proportion of the rain. Second, the grass cover protected the soil surface from being struck directly by the rain and hence minimized the formation of a less permeable crust on the slope surface (Hillel, 1998). Indeed, after the rainfall events, a substantial amount of sediments was observed in the water collection

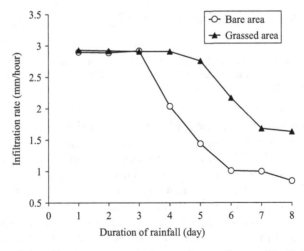

Figure 4.16 Comparison of the infiltration rates in the bare and grassed areas. (Ng, C. W. W. and Zhan, L. T., *Soils Found.*, 47, 207–217, 2007.)

trench at the toe of the slope in the bare area but not in the grassed area. This further showed that grass could prevent the breakdown of soil aggregates and their detachment from the soil mass by raindrops. Third, the grass on the slope surface increased surface roughness. Last, the different soil moisture conditions between the two test areas meant that different levels of initial water permeability likely also contributed to the differences in the infiltration rate.

It was interesting to find that the infiltration characteristics observed in this field study differed from those found in the *in-situ* double-ring infiltration tests conducted at the HKUST Eco-Park (see Figure 4.5) in which the infiltration rate in the grassed soil was lower than that in the bare soil. The observed difference between the two field studies may be attributed to the difference in the soil type tested. The soil tested at the HKUST Eco-Park was a non-expansive, low-plasticity well-graded sand with silt, and no desiccation cracks were identified at the site. Hence, the soil infiltrability of the bare ground was less likely to be reduced by the 'crack-filling' mechanism as was likely the case with the expansive clay at the Hubei site. The desiccation cracks created by excessive drying of highly plastic expansive clay at the Hubei site enhanced the soil infiltrability. Indeed, rainfall infiltration in an unsaturated expansive slope is much more complex than what the one-dimensional infiltration theory predicts as the latter ignores possible preferential flows along cracks and fissures. Water permeability in unsaturated expansive soil is determined by the extent of open cracks and fissures. The comparison of the test results between the two sites suggests that vegetation alone does not always increase or decrease the soil infiltrability. The soil type also plays a significant role in soil–vegetation interaction, its effects on the soil surface condition and hence the soil hydrological responses.

4.3.2 Grass effects on soil pore-water pressure

Figure 4.17 compares the measured variations in PWP with time during the simulated rainfall. Just after rainfall commenced, negative PWP (matric suction) was measured in both areas. The values measured in the grassed area, especially at a depth of 1.6 m, were larger than those measured in the bare area, as expected, owing to grass evapotranspiration (ET) prior to the rainfall simulation. Suction did not decrease in the grassed area until after 2.5–3 days of rainfall, but it dropped in the bare area after just 1.5–2 days, with the exception of suction at a depth of 1.6 m in the middle of the slope (i.e., R2-T-1.6). The delay duration was generally consistent with the onset of surface runoff in both areas. After the onset of runoff, the increase in PWP (or the drop in suction) in the grassed area was more gradual than that in the bare area, especially in the upper part of the slope. At the end of the rainfall, positive PWP values of around 10 kPa were recorded in both areas, and the values were close between the two areas.

Figure 4.18 compares the distributions of PWP with depth in the bare and grassed areas. Before rainfall, suction within the top 1 m of soil in the bare area was significantly higher than that in the grassed area in both sections R1 and R2 (see Figure 4.18a and b). The high suctions within the top 1 m of soil in the bare area, in particular in the upper part of the slope, may be related to the excessive evaporation enhanced by the wide-open cracks near the ground surface. Another reason for the higher soil suction may be attributed to the difference in the initial groundwater conditions prior to the artificial rainfalls between the two areas. Before the rainfall simulation, the bare area was protected with a plastic membrane from rainfall infiltration, whereas the grassed area was exposed to natural precipitation from 1 June to 18 September 2001, during which the total precipitation was about 250 mm. Thus, the higher suction found in the bare area likely led to a lower water permeability but a higher hydraulic gradient in the upper soil layer, which may also have contributed to the observed difference in the infiltration rate between the two areas (see Figure 4.16).

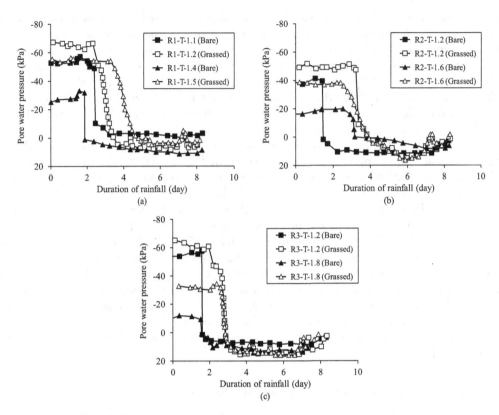

Figure 4.17 Pore-water pressure responses at sections (a) R1; (b) R2 and (c) R3. (Ng, C. W. W. and Zhan, L. T., *Soils Found.*, 47, 207–217, 2007.)

The trend below a depth of 1 m was the opposite of that above. The initial suction in the grassed area was higher than that in the bare area because of ET. Root water uptake and hence soil drying induced suction in the soil near the root system, as was also observed in the laboratory tests discussed in Chapter 2. The field test results shown in Figure 4.18 suggest that the depth of influence of suction due to ET can be four times the grass root depth. This explains why suction at a depth of 1.2 m or more showed a more significant change in the grassed area than in the bare area.

The higher initial suction observed at depths greater than 1 m in the grassed area in all three sections of the slope might suggest that the open cracks in the grassed area extended much deeper than those in the bare area. Note that the depth of open cracks depends on the magnitude of soil suction (Fredlund and Rahardjo, 1993; Ng and Menzies, 2007). Based on linear elastic fracture mechanics, Morris et al. (1994) deduce that the depth of cracks is proportional to the depth of groundwater level and suction, which is assumed to decrease linearly below the ground surface. As intact expansive clay has a low water permeability, rainwater would first infiltrate the deeper soil layer through the open cracks before rising from the bottom of the cracks. This may explain why the PWP responses measured in the grassed area, especially at shallow depths, showed a longer delay than those observed in the bare area (see Figure 4.17).

In the upper and middle portions (i.e., sections R1 and R2) of the slope, the PWP profiles after 1 week of rainfall in the grassed area were not significantly different from those in the bare area. In the lower part of the slope (i.e., section R3), the positive values of PWP in the

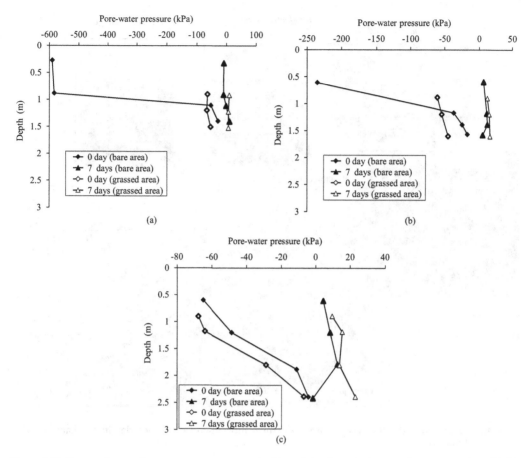

Figure 4.18 Comparisons of pore-water pressure profiles of the bare and grassed areas: sections (a) R1, (b) R2 and (c) R3. (Ng, C. W. W. and Zhan, L. T., *Soils Found.*, 47, 207–217, 2007.)

grassed area, especially at a depth of 2.5 m, were higher than those in the bare area. The higher positive PWP values were likely related to the presence of more significant perched groundwater below the grassed ground surface (Zhan, 2003). Indeed, at the end of the rainfall event, excessive free water was detected at the depth of 2.5 m when samples were collected with an auger from the lower part of the grassed slope (section R3).

4.4 CASE STUDY 3: A NATURAL SAPROLITIC HILLSLOPE IN HONG KONG (VERBATIM EXTRACT FROM LEUNG AND NG, 2013A, 2016)

This field case study was conducted at a natural saprolitic hillslope in Tung Chung, Lantau Island, Hong Kong, where a slow-moving landslide body was identified. The slope had an average angle of 30° and was located in the vicinity of the North Lantau Highway – the primary access to the Hong Kong International Airport (Figure 4.19). The slope was densely vegetated. The most abundant species were two woody species, *Rhodomyrtus tomentosa* and *Baeckea frutescens*, and a fern species, *Dicranopteris pedata*, which are all commonly

Figure 4.19 Overview of the research slope and the three most abundant species identified in Lantau Island, Hong Kong.

found in Hong Kong and many parts of Asia including Malaysia, India, and Vietnam (Corlett et al., 2000). The fern *Dicranopteris pedata* was found to be the most dominant and it covered almost the entire slope surface. The height of vegetation was less than 3 m in general and no canopy was formed in the study area. Based on the species identified, the slope can be described as a typical shrubland.

Figure 4.20 shows the geological profile along the longitudinal direction. The slope comprises four soil strata. In the top 2–3 m of the slope, colluvial deposits, which consisted of silty clay mixed with cobbles made of decomposed tuff, were encountered. Plant rootlets extended to an average depth of 1.2 m. Below the colluvium stratum, a thick layer of completely decomposed coarse ash tuff (CDT) (about 10 m) overlay a 3 m thick layer of highly decomposed coarse ash tuff. Both colluvium and CDT are classified as inorganic silty clay of low to medium plasticity (ASTM, 2011). Below a depth of 10 m, rock materials, including moderately to slightly decomposed coarse ash tuff, were encountered. The initial groundwater table (before the start of the field monitoring) was identified at a depth of about 11 m and closely followed the rock head profile. The mechanical and hydraulic properties of the colluvium and CDT are reported in Ng et al. (2011), Leung et al. (2011), Leung and Ng (2013b, 2016).

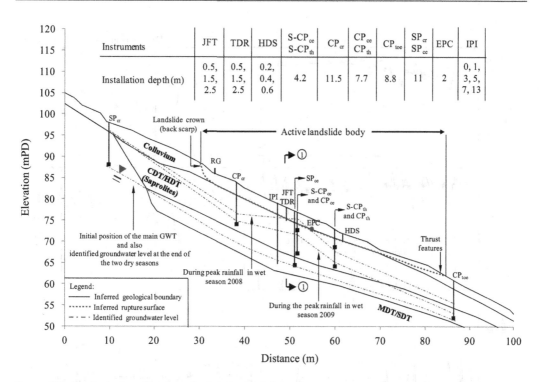

Figure 4.20 Geological profile along the longitudinal direction.

Figure 4.21a–d shows the daily climatic variation of the study area during the monitoring period, which consisted of a wet season from April to September 2008 followed by a dry season from October 2008 to March 2009. The site experienced typical subtropical climate. The air temperature increased (up to 30°C) during summer in the wet season and then decreased (down to 15°C) in winter during the first 3 months of the dry season. For the entire wet season, the monthly average relative humidity (RH) remained fairly constant at about 85%. During the rainy period between May and July 2008, the Hong Kong Observatory issued black rainstorm signals several times (for rainstorms whose intensity exceeds 70 mm/h). The minimum RH dropped to 40% in January 2009. Wind speed fluctuated widely without any observable trend or relationship with season. The typhoon signal no. 8 (for sustained wind speeds of 63–117 km/h and gusts exceeding 180 km/h), together with the black rainstorm signal, had been issued by the Hong Kong Observatory on 7 June 2008 during the rainy period. While there was a substantial increase in radiation from 10 to 20 MJ/m²/d from April to July 2008, measured values were fairly constant at about 15 MJ/m²/d for the entire dry season.

To assist the interpretation of the effects of vegetation on the hydrogeological responses of the research slope, potential evapotranspiration (*PET*; mm/d) was estimated using the Penman-Monteith equation (Allen et al., 1998) as follows:

$$PET = \frac{0.408\Delta(R_n - G) + \gamma \dfrac{900}{T_{air} + 273} u(e_s - e_a)}{\Delta + \gamma(1 + 0.34u)} K_c \tag{4.1}$$

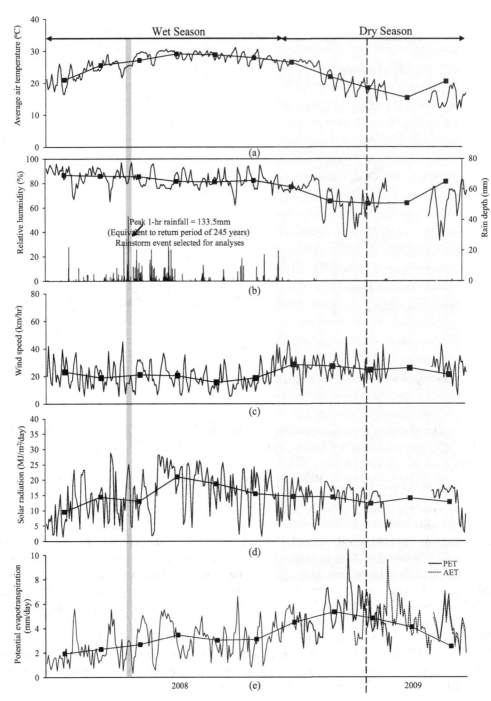

Figure 4.21 Climate of the study area: (a) air temperature, (b) air relative humidity and rain depth, (c) wind speed, (d) solar radiation and (e) calculated actual and potential evapotranspiration.

where *PET* is the amount of soil moisture extracted by plants when pore water is readily available; Δ is the slope of the vapour pressure curve (kPa/°C); R_n is the net radiation intercepted by plant leaves (J/m²/d); G is the soil heat flux density (J/m²/d) (assumed to be negligible, owing to the small magnitude relative to R_n; Allen et al., 1998); γ is the psychometric constant (kPa/°C); T_{air} is the air temperature (°C); u is the wind speed (m/s); e_s is the saturated vapour pressure (kPa); e_a is the actual vapour pressure (kPa); and K_c is the crop factor (taken to be 1.0 for general shrub species found in abundance at the site; Allen et al., 1998). As shown in Figure 4.21e, *PET* was relatively low during the rainy period in the wet season, fluctuating between 0 and 5 mm/d. An increase in *PET* was estimated during the first 2 months of the dry season in winter and the peak *PET* reached 10 mm/d. In the subsequent spring, the estimated *PET* then dropped owing to the substantial increase in air RH.

In order to identify plant hydrological effects on slope hydrology, the vegetated slope was heavily instrumented (Figure 4.20) with heat dissipation sensors (HDSs) and JFTs for measuring matric suction and time domain reflectometers for measuring the VWC, all in the top 2.5 m of the slope. In addition, piezometers (CPs) and standpipes (SPs) were installed to monitor (1) any formation of a perched groundwater table (GWT) near the colluvium-CDT interface at shallow depths during rainfalls and (2) fluctuation of the main GWT at greater depths. These sensors were distributed either near the central part (denoted as the subscript 'ce' in the figure) or the thrust feature (denoted as 'th') of the landslide body.

4.4.1 Plant-induced changes in soil hydrology

Figures 4.22a and b show the variations in PWP and VWC measured at the central portion of the landslide body, respectively. During the rainy period from May to July 2008, PWP at a depth of 0.5 m always exceeded −16 kPa when rainfalls with intensity higher than 20 mm/h occurred. On the contrary, PWP at the larger depths of 1.5 and 2.5 m was higher and positive and varied between 15 and 25 kPa. In response to the substantial increases in PWP during the rainy period, measured VWCs at all depths increased. The measured changes in VWC in CDT at the depth of 2.5 m during most rainfall events varied between 26% and 36% and were much larger than the responses observed in colluvium at the depths of 0.5 and 1.5 m. This may be because CDT has a lower water retention capability than colluvium when they are subjected to the same given change in PWP (Ng et al., 2011). During the no-rain period in August 2008, substantial reductions in PWP (i.e., recovery of suction) were observed within the root zone at a depth of 0.2 m (up to 150 kPa) owing to root water uptake. Considerable recovery of matric suction was occasionally recorded during the no-rain periods in August and September 2008, but the magnitudes were lower (i.e., up to 60 kPa only). This was expected as the durations of the no-rain periods in these months were shorter than those in August 2008.

During the dry season from October 2008 to February 2009, no rain was recorded within the study area. Matric suction increased at all depths, particularly within the root zone at the depths of 0.2, 0.4, and 0.6 m. At the larger depths of 1.5 and 2.5 m, on the contrary, the measured rate of change in PWP was less because the active root zone was much further away. After monitoring for 2 months from October to December 2008, the measured rates of increase in matric suction within the root zone at the depths of 0.2, 0.4, and 0.6 m reduced substantially. Suction at these depths reached the steady-state values of 190, 160, and 165 kPa for the rest of the dry season. A similar reduction in the rate of change in VWC was also found at a depth of 0.5 m on December 2008. When the VWC dropped below 27%, the rate of change in VWC fell considerably before staying fairly constant. In order to interpret the observed hydrological responses within the root zone in relation to soil–plant–water interaction, a simplified water balance calculation was performed. It was assumed that the loss of soil moisture within the root zone was entirely due to plant ET (i.e., there was no recharge from

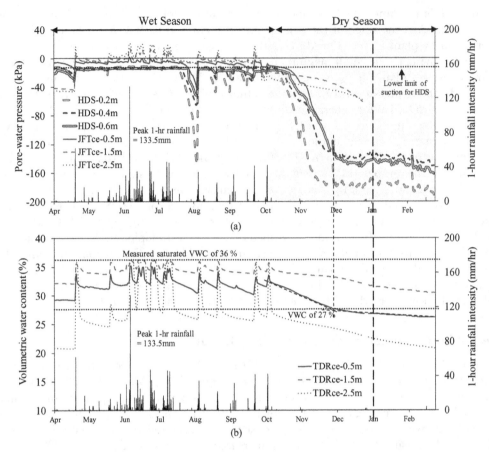

Figure 4.22 Seasonal variations in (a) pore-water pressure and (b) volumetric water content with time under a wet season followed by a dry season.

soil outside the root zone) during the dry season. This assumption was later justified when investigating the distribution of PWP with depth, as will be discussed in the next section. To account for the effects of matric suction (i.e., water stress) on plant root water uptake, the following piecewise linear equations proposed by Feddes et al. (1978) may be used, if plant roots at any depth within the root zone are idealized to have the same ability to extract water:

$$\frac{AT}{PT} \approx \frac{AET}{PET} = \begin{cases} 1 & \text{for} & \psi < \psi_d \\ \dfrac{\psi_w - \psi}{\psi_w - \psi_d} & \text{for} & \psi_w \leq \psi \leq \psi_d \\ 0 & \text{for} & \psi_w < \psi \end{cases} \quad (4.2)$$

where AT is the actual transpiration (mm/d); PT is the potential transpiration (mm/d); ψ_d is the matric suction (kPa) beyond which plant roots have increasing difficulties in extracting moisture from soil; and ψ_w is the matric suction (kPa) corresponding to the wilting point at which plant roots could no longer extract moisture from soil. The ratio of AET to PET was approximated by the ratio of AT to PT. This approximation may be reasonable because evaporation from the bare soil surface was rather limited when the slope surface was densely vegetated (see Figure 4.19). It should be noted that both ψ_d and ψ_w are strongly dependent on soil type, plant species, climate conditions, and their combinations. As a first approximation, typical values of 40 and 1500 kPa were taken for ψ_d and ψ_w (Feddes et al., 1978), respectively. For a given root zone with an average depth of 1.2 m, the water balance calculation was made at the mid-depth at 0.6 m and therefore matric suction measurements recorded by the HDS installed at that particular depth were used for calculation. The calculated variation in AET with time in the dry season is depicted in Figure 4.21e. It can be seen that there was less AET than PET throughout the dry season because the building up of water stress upon soil drying restricted plant root water uptake. However, the difference did not exceed 10%. This means that plants transpired almost at their potential rates (i.e., PET), which were mainly driven by the climatic variation at the site (refer to Eq. 4.2). Based on the water balance calculation, the observed reductions in the rates of change in both matric suction and VWC on December 2008 in Figure 4.22 may be more likely to be attributed to the decrease in AET mainly because of the increase in air RH (see Figure 4.21b).

4.4.2 Plant effects on slope hydrological responses

Figure 4.23 shows the seasonal variations of PWP profiles at the central portion of the landslide body (i.e., across Section 1–1; refer to Figure 4.20). A hydrostatic line representing the initial elevation of the main GWT at a depth of 11 m is also depicted for reference. Initially before the wet season (April 2008), the measured PWP in the top 2.5 m of the soil was always higher than the hydrostatic values, indicating net downward water flow. When rainstorms with a peak intensity of 133.5 mm/d occurred in June 2008, positive PWP was recorded at all depths. Based on the monitoring results from CPs and SPs installed at the central portion of the landslide body, it was revealed that the GWT during this rainstorm event occurred at a depth of about 0.4 m. If a hydrostatic line representing this GWT is shown together with the PWP measurements, it becomes evident that the PWP profile in the top 4 m of the soil was almost hydrostatic. In Figure 4.20, monitoring results from all CPs and SPs showed that during the rainstorm event, the measured GWT in the test slope rose by about 6 m generally. However, a very substantial rise in the GWT of about 10 m was recorded at the central portion of the landslide body (see also Figure 4.20), which then led to the observed building-up of positive PWP in Figure 4.23. This was identified to have caused instability of the slope, which exhibited a considerably large

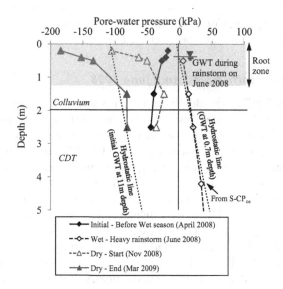

Figure 4.23 Seasonal distributions of pore-water pressure with depth.

down-slope displacement of up to 40 mm and development of rupture surfaces at a depth of about 5 m, leading to deep-seated mode of deformation (Leung and Ng, 2013b).

At the end of the wet season in November 2008, PWP in the top 3 m of the soil dropped substantially below zero, and the GWT returned to the initial elevation of 11 m (see Figure 4.23). Despite the decrease, the PWP was still higher than the hydrostatic value, indicating net downward water flow upon rainfall infiltration. After 4 months of plant ET during the dry season, the measured PWP in colluvium was lower than the hydrostatic value, while the PWP profile in CDT was close to the hydrostatic line. This meant that there was net upward water flow in colluvium under the effects of plant ET, whereas net downward seepage occurred in CDT at greater depths because of gravity. It may thus be determined that the depth of influence of suction during the four-month ET process was shallower than 2 m (i.e., about 200% of the plant root depth).

For the given PWP profile at the end of the dry season in March 2009, it can be estimated that the hydraulic gradient creating upward water flow in colluvium was up to 70 between the depths of 0.2 m (suction of 190 kPa) and 1.5 m (suction of 80 kPa). Although the hydraulic gradient was quite large, the calculated Darcian flow velocity was minimal when the water permeability was taken to be 1×10^{-9} m/s at the average matric suction of 135 kPa (refer to Ng et al., 2011). It was thus anticipated that upward water recharge of the root zone from soil at larger depths was negligible in the shallow ground. This validates the assumption made in the above soil water balance calculation (Eq. 4.2) that plant ET was the main reason for the loss of soil moisture in the root zone.

4.4.3 Transpiration effects on the stress-deformation characteristic of the slope

It is evident that vegetation plays a significant role in slope hydrology by changing the soil moisture and matric suction regime, especially at shallow depths where the landslide body is active. According to unsaturated soil mechanics (Fredlund and Rahardjo, 1993; Ng and Menizes, 2007), changes in matric suction, predominantly induced by plants in this field case, would induce changes in soil volume and stress state. In order to investigate the effects of transpiration-induced suction on the stress-deformation of the field soil and hence the hydro-mechanical behaviour of the landslide body, earth pressure cells and in-place inclinometers were installed in the slope (see Figure 4.20) to measure total horizontal pressure (σ_D) and horizontal displacement, respectively.

4.4.3.1 During the rainstorm from 5 to 9 June 2008

Figure 4.24 correlates PWP at the depths of 0.5, 1.5 and 2.5 m with downslope displacements at the depths of 0, 1 and 3 m, respectively. During the rainfall event on 6 June 2008, a noticeable increase in the horizontal displacement of the slope was recorded when PWP at the depths of 0.5, 1.5 and 2.5 m reached 2, 11 and 15 kPa, respectively. As the slope displaced towards the downslope direction during the storm, the peak positive PWP at all three depths, however, remained almost unchanged. This suggests that the increase in the positive PWP at shallow depths (the top 3 m) could not have been the major reason behind the large downslope displacements. Horizontal displacement profiles measured from 5 to 9 June 2008 are compared in Figure 4.25. During the small rainfall occurring on 6 June, the displacement profile exhibited a cantilever mode of deformation, with the peak change in displacement occurring at the slope surface. After the storm, significantly large downslope displacements were recorded at all depths. This indicates that the depth of influence of slope displacement due to the storm was deeper than 5 m. To determine the depth of influence,

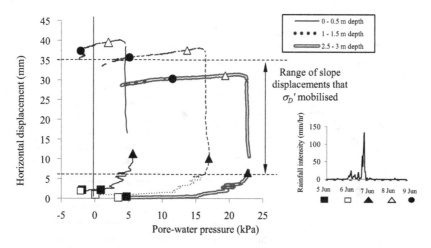

Figure 4.24 Relationships between pore water pressure and total horizontal displacement during the storm event from 5th to 9th Jun 2008.

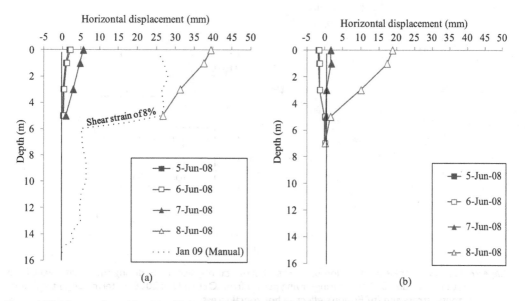

Figure 4.25 Measured total horizontal displacement profiles during the storm event (a) at the central portion and (b) near the thrust features of the landslide body.

a manual inclinometer survey was conducted to a depth of up to 15 m in January 2009. The slope at the location surveyed exhibited a deep-seated mode of movement. A remarkably large displacement of 20 mm was recorded within the stratum from 5.5 to 6 m deep, equivalent to an average shear strain of 8%. Considering the fact that the slope displacement profiles before (7 June 2008) and after (8 June 2008) the storm were almost parallel, the landslide body had likely undergone a translational type of downslope movement along rupture surfaces developed and/or re-activated at the depths of 5.5–6 m. Such translational movement of the top 5 m of the sliding mass may explain why the peak positive PWP did not correspond to the peak slope displacements in the top 3 m of the slope (Figure 4.24).

Figure 4.26 shows the process of stress mobilisation at a depth of 2 m during slope displacement from 5 to 9 June 2008. Since positive PWP and saturated VWC were recorded at a

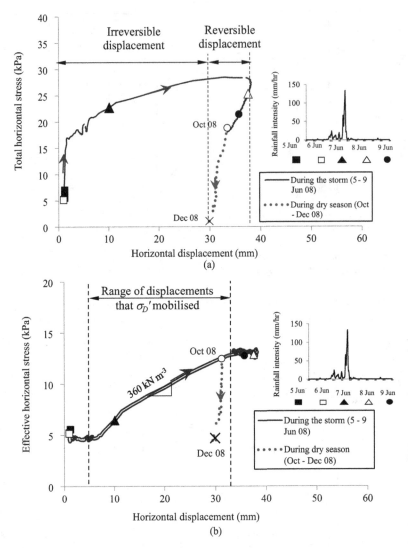

Figure 4.26 Observed stress mobilisation upon total horizontal displacements during the storm event from 5th to 9th Jun 2008 and during drying period from Oct to Dec 2008, in terms of (a) total horizontal stress and (b) Bishop's effective horizontal stress.

depth of 2.5 m during the storm event, effective horizontal stress σ_D' can be determined from the difference between total horizontal stress σ_D and positive PWP, as a first approximation. Initially, at zero slope displacement, σ_D and σ_D' were both 5 kPa and they were attributed to the stress mobilisation during the previous rainfall (with an intensity of up to 60 mm/h) that occurred in April 2008. A displacement of 5 mm during the small rainfall events on 6 June 2008 caused a rather stiff response of σ_D, but σ_D' was not mobilised. This was because the increase in σ_D was mostly attributed to the increase in positive PWP. As the slope shifted a further 30 mm during the storm, a 350% increase in σ_D' (from 4 to 14 kPa) was recorded. By Rankine theory (which was modified to consider the sloping ground condition) and using the effective strength parameters (c' of 7.4 kPa and ϕ' of 33°; see Leung and Ng, 2016), the peak σ_D' was estimated to have been mobilised 40% of Bishop's effective passive stress (38 kPa) of CDT. The mobilisation of σ_D' due to the horizontal slope displacement might be analogous to a horizontal subgrade reaction problem. By first determining the gradient of the linear portion of the σ_D' curve, it may be deduced that the coefficient of the horizontal subgrade reaction, η_h, of CDT was 360 kN/m³. This value is close to the lower range of η_h (350–700 kN/m³) of clayey soil (Tomlinson and Woodward, 2014). The range of displacements (5–35 mm) that mobilised σ_D' corresponds to the constant range of positive PWP. The interrelationship between PWP, σ_D' and displacement suggest that during translational sliding (Figure 4.25), the top 5 m of the sliding mass may have been substantially deformed and caused stress mobilisation at a depth of 2 m. After the storm ceased on 8 June, the mobilised σ_D' remained almost unchanged, as the landslide body rebounded slightly (<5 mm) towards the upslope direction (Figure 4.26). This was because the decrease in σ_D' during this period was almost identical to the reduction in positive PWP. Similar slope movements were also correlated with measured lateral earth pressure in expansive clay and reported by Ng et al. (2003). Consistent observations were found.

4.4.3.2 During the dry season

Figure 4.27 relates horizontal displacements recorded at the depths of 0, 1 and 3 m to plant-induced suction measured at the depths of 0.5, 1.5 and 2.5 m, respectively. As suction increased from 10 to 90 kPa during the drying period, upslope displacement occurred at all depths but at a decreasing rate. Similar upslope movements were also recorded in expansive clay slope (Ng et al., 2003). The void ratio of a soil specimen of the same type in the field loaded with the net stresses of 0 and 40 kPa in the laboratory (Leung and Ng, 2016) reduced at very similar rates to those measured in the field at the depths of 0.5 and 2.5 m, respectively. For the same given increase in suction from 10 to 90 kPa, the increase in upslope displacement (6.3% at the ground surface and 4.3% at a depth of 3 m) in the field was close to the decrease in the void ratio (6.1% at 0 kPa and 4.1% at 40 kPa of net stress) in the laboratory. The close variation suggests that the plant-induced suction, hence shrinkage, was the most likely reason for the upslope rebound in the dry season. It is worth noting that the upslope rebound (by about 10 mm in 5 months) was much smaller and took much longer than the downslope displacement that occurred during the storm (by more than 25 mm in 1 day). Such a large contrast in the rate of slope movement between the wet and dry seasons is the primary reason for the net downslope movement after a year of monitoring. Ratchetting upslope and downslope movements due to seasonal suction changes were also reported in expansive clay slope (Ng et al., 2003).

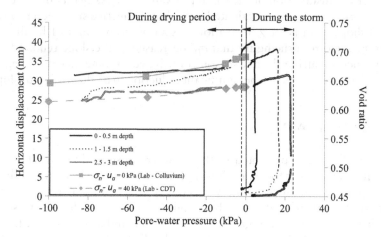

Figure 4.27 Relationships between pore water pressure and total horizontal displacement during the drying period from Oct to Dec 2008.

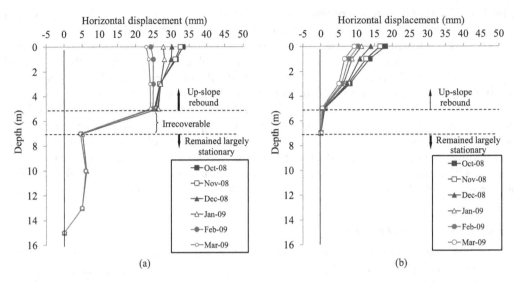

Figure 4.28 Measured total horizontal displacement profiles during the drying period (a) at the central portion and (b) near the thrust features of the landslide body.

Figure 4.28 shows the monthly displacement profiles during the drying period between October 2008 and March 2009. The displacements in the top 5 m of soil reduced, whereas the soil below 7 m was largely stationary. The observed decreases in displacement indicate that the soil shifted up the slope during the drying period. The large displacement at 5.5–6 m previously recorded was almost irreversible. The reduction in displacement was smaller at larger depths. This deformation mode may be the consequence of the decrease in shrinkability of soil with increasing depth. Indeed, soil subjected to a higher vertical load (equivalent to overburden pressure in the field) has a lower reduction in the void ratio owing to a higher soil stiffness (Leung and Ng, 2016). As the vertical load increased from 0 to 80 kPa (equivalent to an increase in depth from 0 to 4 m), the decrease in void ratio reduced from 7.2% to 3.5%.

As the slope rebounded by 5 mm, it can be seen from Figure 4.26 that σ_D' built up during the previous storm reduced substantially from 20 to 0 kPa (i.e., returning to the initial value of σ_D after the installation of the earth pressure cells). As plants induced suction during the drying period, σ_D' may be evaluated by Bishop's effective stress equation considering the value of Bishop's parameter χ in relation to suction according to Khalili and Khabbaz (1998). It can be seen from the figure that σ_D' decreased as the slope rebounded. Although the increase in suction during the dry season contributed to some increase in σ_D', the overall reduction in σ_D' was attributed to the significant reduction in σ_D during the upslope rebound.

4.5 CHAPTER SUMMARY

The three sets of field data presented in this chapter consistently show that plants (grass and trees) significantly affect the water infiltration rate and matric suction. The infiltration rate of non-expansive soil (such as the well-graded sand with silt tested in case study 1) with vegetation was half of that of the bare soil. Plant-induced changes in soil infiltrability result in suction preserved in vegetated slopes during and after rainfall. In contrast, the infiltration rate of expansive soil (such as the highly plastic clay tested in case study 2) with vegetation was higher than that of the bare soil, owing to the formation of open desiccation cracks on the soil surface via ET. Apart from the soil type, vegetation management also plays an important role in the infiltration rate and water permeability. Growing trees too close together (i.e., when spacing is equal to the lateral spread of the tree root system) introduces intra-species competition for water, causing root decay and formation of macro-pores and hence preferential flows. On the contrary, growing the same trees at a wider spacing (i.e., one to two times the root lateral spread) lowered the water permeability of the soil. Thus, vegetation alone does not always increase or decrease soil infiltrability. The latter is a function of multiple factors in relation to soil type and vegetation management, both of which play a significant role in soil–plant–water interaction and its effects on soil hydrological responses. More research is needed to determine the optimal planting space and mixed planting criterion for slope stabilisation.

Another observation consistently found in all three field studies was that plant-induced suction could be preserved mainly at greater depths (below the root zone), whereas suctions at shallower depths (within the root zone) largely disappeared after rainfall. This implies that plant transpiration and root-induced changes in soil hydraulic properties (soil water retention and water permeability) are more important for stabilising shallow soil (~1–2 m, where the translational type of failure is typically of major concern) than mechanical root reinforcement, which is more relevant to shallow slope stabilisation.

Owing to plant-induced changes in soil hydrology, soil volume and stress also change accordingly, affecting the mechanical behaviour of soil slopes. The field monitoring results from case study 3 revealed upslope rebound movement during a drying season, when matric suction was developed by ET, reversing the downslope movement, owing to a previous extreme storm event, albeit by no more than 25% (i.e., about 10 mm). For a given increase in induced suction, the increase in upslope displacement (4.3%–6.3%) in the field was close to the decrease in the void ratio (4.1%–6.1%) of the soil tested at similar levels of overburden stress in the laboratory. This clearly suggests that the upslope rebound was attributed to soil shrinkage due to suction recovery by ET. Since the soil shrinkability reduced with an increase in overburden stress, a smaller upslope displacement was observed at greater depths. This thus led to a cantilever shape of the upslope displacement profile.

Chapter 5

Theoretical modelling of plant hydrological effects on matric suction and slope stability

5.1 INTRODUCTION

Reliable prediction of transpiration-induced soil moisture and changes in soil suction is important for transient seepage analysis in unsaturated vegetated soils and hence for estimating soil hydrological changes and determining slope stability. There are two broad approaches to modelling transpiration and root water uptake, namely macroscopic and microscopic approaches. The former integrates a so-called macroscopic sink term into the Darcy–Richards equation (Feddes et al., 1976; Hopmans and Bristow, 2002), and this sink term is a function of climate and of the ability of plant roots to take up water (Casaroli et al., 2010; Fatahi et al., 2010; Świtała et al., 2018). The latter approach views the soil–plant–water continuum as analogous to an electric circuit (Roose and Fowler, 2004; Janott et al., 2011) and uses a number of resistance (or conductivity) terms to describe the ease of water flow in different components of a plant such as a leaf, the stem, and the root. Because of its simplicity and the relative ease in calibrating the input parameters, the macroscopic approach has been more commonly used to analyse plant effects on the engineering performance of earthen structures, including slopes and landfill covers (Fatahi et al., 2010; Briggs et al., 2016; Ng et al., 2018c). Although this modelling approach appears to be appealing, it has a number of major limitations. For example, it ignores the effects of (1) the root architecture on the magnitude and distribution of plant-induced suction and (2) root-induced changes in soil hydraulic properties, including soil water retention ability and water permeability. All these parameters have significant effects on the soil hydrology (see Chapter 2). This chapter presents advanced theoretical analyses that capture these previously ignored hydrological effects of plants on changes in soil suction and consequently the stability of unsaturated vegetated soil slopes. Closed-formed, analytical steady- and transient-state solutions are derived to estimate the magnitude and distribution of plant-induced soil suction and hence the stability of an infinite vegetated slope with different root architectures.

5.2 PLANT TRANSPIRATION-INDUCED CHANGES IN MATRIC SUCTION AND SLOPE STABILITY (NG ET AL., 2015; LIU ET AL., 2016)

5.2.1 Governing equations

Figure 5.1 shows an infinite slope of angle φ. It is assumed that the groundwater table is located at the bottom of the slope, and the equipotential lines of pore-water pressure are parallel to the slope surface. Thus, the water flow can be treated as a one-dimensional flow perpendicular to the slope at both steady and transient states. For a vegetated slope, the governing equation can be derived according to a procedure similar to that used by Zhan et al. (2013) but with a sink term $S(z')$ added to consider root water uptake as follows:

$$\frac{\partial}{\partial z'}\left(k\frac{\partial h}{\partial z'}\right)+\frac{\partial k}{\partial z'}\cos\varphi - S(z')H(z'-L_1') = \frac{\partial\theta}{\partial t} \tag{5.1}$$

where h is the pressure head; k is the water permeability; θ is the volumetric water content; t is the elapsed time; z' is the coordinate perpendicular to the slope surface, as shown in the figure; and $H(z'-L_1')$ is the Heaviside function (Polyanin, 2002), defined as:

$$H(z'-L_1') = \begin{cases} 1 & L_1' \le z' \le (L_1'+L_2') \quad \text{inside the root zone} \\ 0 & 0 \le z' \le L_1' \quad \text{outside the root zone} \end{cases} \tag{5.2}$$

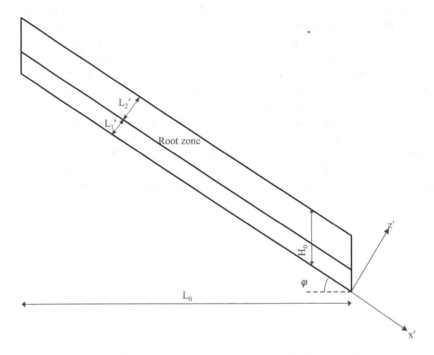

Figure 5.1 A schematic diagram of an infinite vegetated slope and definition of variables.

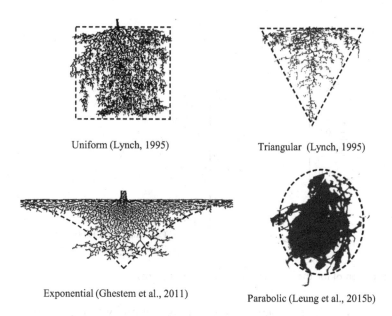

Uniform (Lynch, 1995)

Triangular (Lynch, 1995)

Exponential (Ghestem et al., 2011)

Parabolic (Leung et al., 2015b)

Figure 5.2 Four typical root architectures observed in the literature.

where L_1' and L_2' are the perpendicular depths to the slope surface of the outside root zone and the root depth, respectively. Based on the field and laboratory data reported in the literature (Lynch, 1995; Ghestem et al., 2011; Leung et al., 2015b), the four most common root architectures are uniform, triangular, exponential and parabolic (Figure 5.2). According to these root architectures, a shape function $g(z)$ that describes the ability of plants to take up water at a certain depth, z, can be defined (Figure 5.3). The corresponding sink term can then be expressed as follows:

$$S(z') = \begin{cases} \dfrac{T}{L_2'} & \text{uniform root architecture} \\[2em] \dfrac{2T}{L_2'}\left(\dfrac{z'-L_1'}{L_2'}\right) & \text{triangular root architecture} \\[2em] T\left[\dfrac{\exp(z'-L_1')-1}{\exp(L_2')-L_2'-1}\right] & \text{exponential root architecture} \\[2em] \dfrac{2T}{L_2'}\left[\dfrac{3((z'-L_1')L_2'-(z'-L_1')^2)}{L_2'^2}\right] & \text{parabolic root architecture} \end{cases} \tag{5.3}$$

In Eq. (5.3), it is assumed that roots grow perpendicular to the slope surface, based on the observations made by Ghestem et al. (2011) and Danjon et al. (2008, 2013). Note that for a fair comparison, in Eq. (5.3), the total transpiration rate (T) is the same for the four types of root architecture and can be obtained by integrating $S(z')$ over the entire root area.

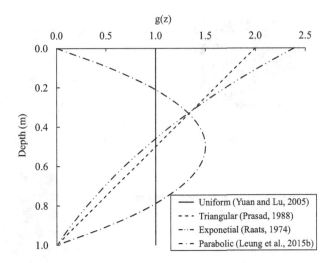

Figure 5.3 Distribution of for the four different root architectures.

In an unsaturated soil, both volumetric water content and water permeability are a function of pressure head h, and they can be expressed according to Gardner (1958) as follows:

$$\theta = \theta_r + (\theta_s - \theta_r)\exp(\alpha h) \tag{5.4}$$

$$k = k_s \exp(\alpha h) \tag{5.5}$$

where k_s is the saturated water permeability of soil; α is the desaturation coefficient of soil; and θ_s and θ_r are the saturated and residual water content, respectively. The effect of hydraulic hysteresis is ignored.

Substituting Eqs. (5.4) and (5.5) into Eq. (5.1) yields Eq. (5.6), a linear partial differential equation, as follows:

$$\frac{\partial^2 k}{\partial z'^2} + \alpha \cos\varphi \frac{\partial k}{\partial z'} - \alpha S(z')H(z'-L_1') = \frac{\alpha(\theta_s - \theta_r)}{k_s}\frac{\partial k}{\partial t} \tag{5.6}$$

For a given set of initial and boundary conditions, both steady and transient state solutions of Eq. (5.1) can be derived. In this study, the following variables are defined:

$$z^* = z'\cos\varphi;\ L^* = L'\cos\varphi;\ L_1^* = L_1'\cos\varphi;\ L_2^* = L_2'\cos\varphi \tag{5.7}$$

$$k^* = k/k_s;\ q_0^* = q_0/k_s;\ q_1^* = q_1/k_s \tag{5.8}$$

where L' is the total perpendicular depth of slope, which is equal to the sum of L_1' and L_2', and k^* is the relative water permeability. q_0 is the surface flux (evaporation or rainfall flux) at steady state, and q_1 is the surface flux at transient state. Both q_0 and q_1 can remain constant or change with time. Substituting Eqs (5.7) and (5.8) into Eq. (5.6) yields

$$\frac{\partial^2 k^*}{\partial z^{*2}} + \alpha \frac{\partial k^*}{\partial z^*} - \frac{\alpha S\left(\dfrac{z^*}{\cos\varphi}\right) H\left(\dfrac{z^*}{\cos\varphi} - \dfrac{L_1^*}{\cos\varphi}\right)}{k_s \cos^2\varphi} = \frac{\alpha(\theta_s - \theta_r)}{k_s \cos^2\varphi} \frac{\partial k^*}{\partial t} \tag{5.9}$$

with the following boundary conditions:

$$k^*\big|_{z^*=0} = \exp(\alpha h_0) \text{ at the bottom boundary} \tag{5.10}$$

where h_0 is the pressure head at the bottom,

$$\left[\frac{\partial k^*}{\partial z^*} + \alpha k^*\right]_{z^*=L^*} = -\alpha q_0^* \text{ at the top boundary at steady state, and} \tag{5.11}$$

$$\left[\frac{\partial k^*}{\partial z^*} + \alpha k^*\right]_{z^*=L^*} = -\alpha q_1^* \text{ at the top boundary at transient state.} \tag{5.12}$$

5.2.2 Steady-state solutions

At steady state, the volumetric water content does not change with time, and the corresponding steady-state equation for Eq. (5.9) is written as

$$\frac{\partial^2 k^*}{\partial z^{*2}} + \alpha \frac{\partial k^*}{\partial z^*} - \frac{\alpha S\left(\dfrac{z^*}{\cos\varphi}\right) H\left(\dfrac{z^*}{\cos\varphi} - \dfrac{L_1^*}{\cos\varphi}\right)}{k_s \cos^2\varphi} = 0, \tag{5.13}$$

with the top boundary condition of:

$$\left[\frac{\partial k^*}{\partial z^*} + \alpha k^*\right]_{z^*=L^*} = -\alpha q_0^*, \tag{5.14}$$

and the bottom boundary condition of:

$$k^*\big|_{z^*=0} = \exp(\alpha h_0). \tag{5.15}$$

Using the same procedure as that proposed by Yuan and Lu (2005), the solution to Eq. (5.13) for both inside (with the sink term) and outside (without the sink term) the root zone with boundary conditions (Eqs [5.14] and [5.15]) can be derived as follows:

$$k_0^* = \exp\left[\alpha\left(h_0 - z^*\right)\right] + q_0\left[\exp\left(-\alpha z^*\right) - 1\right]/k_s$$

$$+ \frac{\alpha}{k_s \cos^2\varphi} \int_0^{L^*} G(z^*, x^*) S\left(\frac{x^*}{\cos\varphi}\right) H\left(\frac{z^*}{\cos\varphi} - \frac{L_1^*}{\cos\varphi}\right) dx^* \tag{5.16}$$

where $G(z^*, x^*)$ is the Green function and is defined as:

$$G(z^*, x^*) = \begin{cases} \exp(-\alpha z^*)\left[1 - \exp(\alpha x^*)\right]/\alpha & 0 \le x^* \le z^* \le L^* \\[2mm] \left[\exp(-\alpha z^*) - 1\right]/\alpha & 0 \le z^* \le x^* \le L^* \end{cases} \tag{5.17}$$

By substituting Eqs (5.3) and (5.17) into Eq. (5.16), the steady-state solutions for the uniform, triangular, exponential and parabolic root architectures can be derived as follows:

1. Uniform root architecture

$$k_s^* = \begin{cases} A + \dfrac{1}{k_s \cos^2 \varphi} \dfrac{T}{L_2'} \left[\exp(-az^*) - 1\right]\left(L^* - L_1^*\right) & \text{outside the root zone} \\[4mm] A + \dfrac{1}{k_s \cos^2 \varphi} \dfrac{T}{L_2'} \left\{ \begin{aligned} &\left[\exp(-az^*) - 1\right]\left(L^* - z^*\right) + \exp(-az^*) \\ &\left[z^* - L_1^* - a^{-1}\exp(az^*) + a^{-1}\exp(aL_1^*)\right] \end{aligned} \right\} & \text{inside the root zone} \end{cases} \tag{5.18}$$

where $A = \exp\left[a\left(h_0 - z^*\right)\right] + q_0\left[\exp(-az^*) - 1\right]/k_s$

2. Triangular root architecture

$$k_s^* = \begin{cases} A + \dfrac{1}{k_s \cos^2 \varphi} \dfrac{2T}{L_2'^2} \left[\exp(-az^*) - 1\right]\left[\dfrac{\left(L^{*2} - L_1^{*2}\right)}{2\cos\varphi} - L_1' L^* + L_1' L_1^*\right] & \text{outside the root zone} \\[4mm] A + \dfrac{a}{k_s \cos^2 \varphi} \dfrac{2T}{L_2'^2} \left\{ \begin{aligned} &\dfrac{1}{a}\left[\exp(-az^*) - 1\right]\left(\dfrac{L^{*2} - z^{*2}}{2\cos\varphi} - L_1' L^* + L_1' z^*\right) \\ &+ \dfrac{\exp(-az^*)}{a}\left[\dfrac{1}{a^2 \cos\varphi}\exp(aL_1^*)\left(aL_1^* - 1\right)\right. \\ &\left. - \dfrac{1}{a^2 \cos\varphi}\exp(az^*)\left(az^* - 1\right) - L_1' z^* + L_1^* L_1'\right] \\ &- \dfrac{L_1^{*2} - z^{*2}}{2\cos\varphi} - \dfrac{L_1'}{a}\exp(aL_1^*) + \dfrac{L_1'}{a}\exp(az^*)\right] \end{aligned} \right\} & \text{inside the root zone} \end{cases} \tag{5.19}$$

3. Exponential root architecture

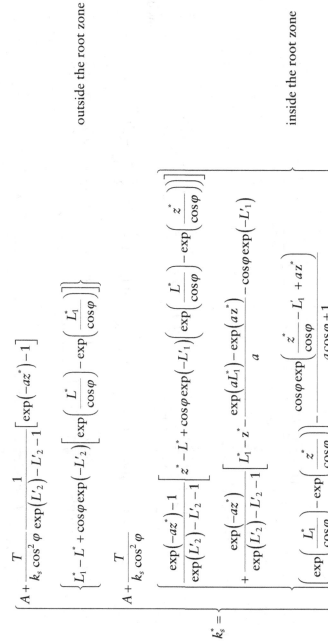

outside the root zone

inside the root zone

$$(5.20)$$

4. Parabolic root architecture

$$
k_s^* =
\begin{cases}
A + \dfrac{1}{k_s \cos^2\varphi}\dfrac{6T}{L_2'^3}\left[\exp(-az^*)-1\right]\left[\dfrac{(L^*-L_1)(2L_1'+L_2')(L^*+L_1)}{2\cos\varphi} - L_1'(L^*-L_1)(L_1'+L_2') - \dfrac{(L^*-L_1)(L^{*2}+L^*L_1+L_1^{*2})}{3\cos^2\varphi}\right] & \text{outside the root zone} \\[4ex]
\end{cases}
$$

$$
\begin{aligned}
A + \frac{1}{k_s \cos^2\varphi}\frac{6T}{L_2'^3}\Bigg[& \left[\exp(-az^*)-1\right]\left[\frac{(L^*+z^*)(2L_1'+L_2')(L^*-z)}{2\cos\varphi} - \frac{(L^*-z)(L^{*2}+L^*z+z^{*2})}{3\cos^2\varphi}\right. \\
& \left. - L_1'(L^*-z)(L_1'+L_2')\right] \\
& + \exp(-az^*)\left[z^{*2}z^*\left(\frac{L_1'}{\cos\varphi} - \frac{z^*}{3\cos^2\varphi}\right) - z^*L_1'^2 + L_1'L_1^{*2} + L_1^{*2}\left(\frac{L_1^*}{3\cos^2\varphi} - \frac{L_1'}{\cos\varphi}\right) - \frac{L_1^{*2}L_2'}{2\cos\varphi}\right] \\
& - \frac{L_2'}{a^2\cos\varphi}\exp(az^*)(az^*-1) + L_1^*L_1'L_2' + \frac{L_2'}{a^2\cos\varphi}\exp(aL_1^*)(aL_1^*-1) + \frac{L_1'^2z^{*2}}{2\cos\varphi}\exp(az^*) - z^*L_1'L_1'L_2' \\
& - \frac{L_2'L_1'}{a}\exp(aL_1^*) + \frac{L_2'L_1'}{a}\exp(az^*) + \frac{\exp(az^*)}{a^3\cos^2\varphi}\left[L_1'^2a^2\cos^2\varphi + 2aL_1'\cos\varphi - 2L_1'z^*a^2\cos\varphi \right. \\
& \left. + z^{*2}a^2 - 2az + 2\right] \\
& - \frac{\exp(aL_1^*)}{a^3\cos^2\varphi}\left(L_1^{*2}a^2 - 2a^2L_1^*L_1'\cos\varphi - 2aL_1^* + L_1'^2a^2\cos^2\varphi + 2aL_1'\cos\varphi + 2\right)\Bigg]
\end{aligned}
\qquad (5.21)
$$

inside the root zone

Using Eqs (5.4) and (5.16), pore water pressure (PWP) u_w can be derived as:

$$u_w = 10\alpha^{-1}\ln\left\{\begin{array}{l}\exp\left[\alpha\left(h_0 - z^*\right)\right] + \dfrac{q_0}{k_s}\left[\exp\left(-\alpha z^*\right) - 1\right] + \dfrac{\alpha}{k_s\cos^2\varphi}\displaystyle\int_0^{L^*}G(z^*,x^*) \\ S\left(\dfrac{x^*}{\cos\varphi}\right)H\left(\dfrac{z^*}{\cos\varphi} - \dfrac{L_1^*}{\cos\varphi}\right)dx^*\end{array}\right\} \quad (5.22)$$

The first term on the right-hand side of Eq. (5.22) represents the hydrostatic PWP distribution. The second term denotes the effects of surface flux on the PWP distribution at steady state. The third term describes the effects of root water uptake.

5.2.3 Transient-state solutions

The steady-state solution k_0^* (Eq. 5.16) is taken as the initial condition for Eq. (5.9) in deriving transient-state solutions. By Laplace transform, the ordinary differential Eq. (5.9) can be derived as:

$$\frac{\partial^2\overline{k^*}}{\partial z^{*2}} + \alpha\frac{\partial\overline{k^*}}{\partial z^*} - s'\frac{\alpha(\theta_S - \theta_r)}{k_s\cos^2\varphi}\overline{k^*} + \frac{\alpha(\theta_S - \theta_r)}{k_s\cos^2\varphi}k^*(z^*,0) - \frac{\alpha S\left(\dfrac{z^*}{\cos\varphi}\right)H\left(\dfrac{z^*}{\cos\varphi} - \dfrac{L_1^*}{\cos\varphi}\right)}{s'k_s\cos^2\varphi} = 0, \quad (5.23)$$

with the top boundary condition of

$$\left[\frac{\partial\overline{k^*}}{\partial z^*} + \alpha\overline{k^*}\right]_{z^*=L^*} = -\alpha\overline{q_1^*} \quad (5.24)$$

and the bottom boundary condition of

$$\overline{k^*}\Big|_{z^*=0} = \exp\left(\alpha h_0\right)/s', \quad (5.25)$$

where s' is the Laplace-transform complex variable; $\overline{k^*} = L(k^*)$; and $\overline{q_1^*} = L(q_1^*)$, where L denotes the Laplace-transform operator.

The analytical solution, Eq. (5.23), is derived using a similar method to that of Yuan and Lu (2005):

$$k^* = k_0^* + 8\frac{\cos^2\varphi}{(\theta_s - \theta_r)}\exp\left[\frac{\alpha\left(L^* - z^*\right)}{2}\right]\sum_{n=1}^{\infty}\frac{\left(\lambda_n^2 + \dfrac{\alpha^2}{4}\right)\sin\left(\lambda_n L^*\right)\sin\left(\lambda_n z^*\right)}{2\alpha + \alpha^2 L^* + 4L^*\lambda_n^2}G(t) \quad (5.26)$$

$$G(t) = \int_0^t\left[q_0 - q_1(\tau)\right]\exp\left[-\frac{\cos^2\varphi}{\theta_s - \theta_r}\left(\lambda_n^2 + \frac{\alpha^2}{4}\right)(t - \tau)\right]d\tau \quad (5.27)$$

where k_0^* is the steady-state solution for different root architectures (Eqs 5.18 through 5.21), and λ_n is the nth positive root of the equation $\sin(\lambda L^*) + (\frac{2\lambda}{\alpha})\cos(\lambda L^*) = 0$. Since the sink term S is independent of time, the transient part in Eq. (5.26) does not depend on root water uptake.

Using the same approach as that for deriving PWP at steady state, the PWP at transient state can be expressed as:

$$u_w = 10\alpha^{-1}\ln\left\{ \begin{array}{c} k_0^* + 8\dfrac{\cos^2\varphi}{(\theta_s-\theta_r)}\exp\left[\dfrac{\alpha\left(L^*-z^*\right)}{2}\right] \\[2em] \displaystyle\sum_{n=1}^{\infty}\dfrac{\left(\lambda_n^2+\dfrac{\alpha^2}{4}\right)\sin\left(\lambda_n L^*\right)\sin\left(\lambda_n z^*\right)}{2\alpha+\alpha^2 L^*+4L^*\lambda_n^2}G(t) \end{array} \right\} \qquad (5.28)$$

The first term in Eq. (5.28) represents the initial PWP distribution, considering root water uptake, which is the solution at the steady state (Eq. 5.22). The second term describes the effects of rainfall on PWP distributions. Both the steady- and transient-state solutions have been verified by comparing them with the predictions made by a finite element software COMSOL. More details can be found in Ng et al. (2015).

5.2.4 Root architecture effects on soil matric suction

5.2.4.1 Effects of root architecture on steady-state PWP

Figure 5.4 shows that the simulated results of the steady-state matric suction (i.e., negative PWP) induced in all four vegetated slopes are all higher than that in the bare slope, as would be expected, because soil moisture is removed via transpiration (c.f. Figures 2.11 and 4.12). Among the four root architectures, the exponential and triangular ones induce the highest matric suction at the slope surface at steady state, followed by the uniform and parabolic ones. This is because the roots near the soil surface are much denser in the exponential and triangular architectures than in the other ones (Figure 5.3). This suggests that plants having a triangular or exponential root architecture can help to reduce PWP (hence water permeability; see Figure 5.4) and also rainfall infiltration.

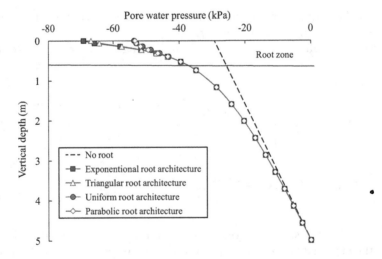

Figure 5.4 Effects of root architecture on steady-state pore-water pressure distributions during drying.

Although there is almost no difference in the steady-state PWP distributions among the four root architectures below the root zone (Figure 5.4), plant root water uptake induces a significant suction at depths below the root zone (compared with that in the bare case). The depth of influence of transpiration-induced suction is about six times the root depth. Indeed, the depth of influence of suction is dependent on both the soil type and the plant type. For example, laboratory test results show that the depth of influence of tree-induced suction in silty sand was 3.5 times the root depth (see Figure 2.11). On the other hand, the grass species in the field tests reported by Ng and Zhan (2007) affected suction in expansive soil up to a depth of 2.5 m.

5.2.4.2 Effects of root architecture on transient-state PWP

Figure 5.5 compares the transient responses of suction distributions in the bare and vegetated slopes during a 24-h rainfall event (equivalent to a 10-year return period). Note that during rainfall, no transpiration was considered. Before rainfall, the suction induced by transpiration in the two vegetated slopes, regardless of the root architectures considered, is higher than that in the bare slope. After rainfall, the suction at shallower depths within the root zone is largely reduced. At the end of the rainfall event, the differences in suction between the bare and vegetated slopes are practically negligible at shallow depths. However, at greater depths below the root zone, for example, at 1–1.5 m, where the translational type of slope failure is of a major concern to slope/geotechnical engineers, the suction preserved in both vegetated slopes is noticeably higher than that in the bare slope (by up to 30%). The major engineering implication here is that the hydrological effect of transpiration-induced suction before rainfall will affect the amount of suction preserved after rainfall. In addition, the hydrological effect of vegetation on slope stabilisation is significant at relatively large depths (1–1.5 m), sometimes below the root zone. At relatively shallow depths where the hydraulic gradient due to rainfall is relatively high (<0.5 m), suction is almost reduced to zero, resulting in only marginal stabilisation effects. Instead, mechanical root reinforcement may be more important for shallow slope stabilisation, as most of the root biomass exists in the top 0.5 m within the root zone.

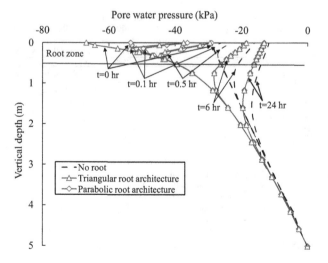

Figure 5.5 Comparison of the transient pore-water pressure distributions of vegetated and bare slopes.

Although the root architecture plays a role in the steady-state suction distribution after drying (see Figure 5.4), its effects are practically negligible under transient rainfall conditions. At the end of rainfall, there is almost no difference in the suction distribution between the exponential and parabolic root architectures (Figure 5.5).

5.2.4.3 Effects of root depth

The triangular and parabolic root architectures with root depths of 0.3, 0.5 and 1.0 m are compared to investigate the effects of root depth on suction distribution in Figure 5.6. The suction near the slope surface increases as root depth increases, regardless of the shape of the root system. At the steady state, the transpiration-induced suction associated with the triangular architecture is as high as 95 kPa for the root depth of 0.3 m, which is almost double than that for the larger root depth of 1 m. Because the transpiration rate is identical in all cases, for a given root architecture, plants having a shallow root depth would have to transpire more water (hence induce higher suction) than deep-rooted plants. It is interesting to see that below 0.5 m, the steady-state PWP distributions are almost the same in all cases, regardless of the root depth and root architecture considered. At this particular depth, the suction is approximately 40 kPa, which corresponds to a soil water permeability of approximately 1×10^{-8} m/s. This has the same order of magnitude as the applied transpiration rate of 6.6 mm/day (i.e., 7.6×10^{-8} m/s). This means that at depths where suction is lower than 40 kPa, the soil water permeability is high enough to reduce the hydraulic gradient created (hence suction induced) by root water uptake. In other words, the ratio of soil water permeability to transpiration rate plays a significant role in how much matric suction is induced by plants.

As far as the design of vegetated landfill covers is concerned, in order to avoid root penetration through the cover system, plants that have small root depths and either a triangular or an exponential root architecture are preferable (Roberts et al., 2006; Hemenway, 2015). Yet, plants with deep roots may be more beneficial for system performance, because deep roots can help the soil recover its water storage capacity by drying the surrounding soil. Engineers must strike a balance when selecting plant species for a landfill cover. In either case, high suction induced by plants at shallow depths could lead to an increased stability of shallow slopes (Liu et al., 2016).

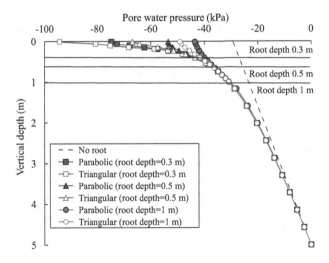

Figure 5.6 Effects of root depth on steady-state pore-water pressure distributions under transpiration.

Moreover, the reduction in soil water permeability due to the increase in transpiration-induced suction could potentially reduce rainfall infiltration and hence reduce the formation of leachate in landfills (Ng et al., 2016b).

5.2.5 Root architecture effects on slope stability

For simplicity, it may be assumed that the shear strength of unsaturated soils obeys the extended Mohr–Coulomb failure criterion (Fredlund and Rahardjo, 1993):

$$\tau_f = c' + (\sigma_n - u_a)\tan\phi' + (u_a - u_w)\tan\phi^b \tag{5.29}$$

where c' is the effective cohesion, $\sigma_n - u_a$ is the net normal stress, $u_a - u_w$ is the matric suction, ϕ' is the effective friction angle and ϕ^b is the angle indicating the rate of increase in shear strength relative to negative PWP. Note that shear strength varies nonlinearly with matric suction (Gan et al., 1988). Considering small matric suction change of less than 100 kPa in the following slope stability analysis and for simplicity, a constant value of ϕ^b is assumed.

Factor of safety (FOS) of a slope is defined as the ratio of available shear strength (i.e., Eq. 5.29) to the driving force of a sliding mass above a pre-defined slip. Based on the force equilibrium parallel to the slope surface, the FOS can be obtained as follows (Zhan et al., 2013):

$$\text{FOS} = \frac{\left(c' - u_w \tan\phi^b\right)}{\left[\gamma_d(H_0 - Z) + \gamma_w \int_Z^{H_0} \theta dZ\right]\sin\varphi\cos\varphi} + \frac{\tan\phi'}{\tan\varphi} \tag{5.30}$$

where γ_d is the dry unit weight of soil; γ_w is the unit weight of water; u_w is the PWP, which can be calculated by the analytical solutions in Eq. (5.28); θ is the volumetric water content; Z is the vertical coordinate with upwards positive, as shown in Figure 5.1; H_0 is the vertical thickness of slope and φ is the slope angle. Note that in this stability calculation, the mechanical contribution of root reinforcement is ignored. Only the contribution of transpiration-induced suction to slope stability is investigated. More detailed analysis considering both the mechanical and hydrological effects of vegetation on slope stability are given in Section 5.3.3.

The infinite vegetated slope considered in this analysis has a slope angle (φ) of 35° and the vertical thickness (H_0) of 5 m (see Figure 5.1). The perpendicular root depth (L_2') is 0.5 m. The soil type considered is completely decomposed granite (CDG, silty sand), which has γ_d of 15 kN/m³, c' of 10 kPa, ϕ' of 38° and ϕ^b of 15°, according to Zhai et al. (2000). Regarding the hydraulic properties, the CDG has k_s of 2.2 × 10⁻⁶ m/s, α of 1.1 m⁻¹, θ_s of 0.45 and θ_r of 0.05 (Chiu, 2001). Three types of slope were considered: bare slope and vegetated slopes with exponential and parabolic root architectures. The root depth in both vegetated slopes was 0.5 m. Before rainfall, steady-state PWP distributions of all slopes were determined by applying a constant transpiration rate of 4.5 mm/d, using the solution expressed in Eq. (5.22). Subsequently, an identical rainfall event of 181 mm/d for 24 h (equivalent to 10 years return period; Lam and Leung, 1995) was applied, and the transient PWP distributions were calculated by the transient-state analytical solutions (Eq. 5.28).

Based on the calculated PWP distributions and using the FOS equation (Eq. 5.30), the FOS profiles along the three slopes (bare slope and two vegetated slopes with parabolic

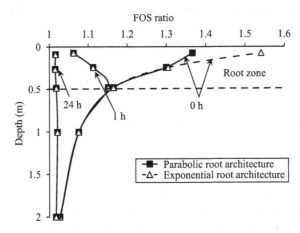

Figure 5.7 The ratio of FOS in vegetated to bare slope under rainfall intensity with 181 mm/day for 24 h. (Liu. H. W. et al., *Comput. Geotech.*, 80, 115–120, 2016.)

and exponential root architectures) were determined. In Figure 5.7, a term, FOS ratio, is defined as the ratio of FOS of a vegetated slope to that of the bare slope. Before rainfall ($t = 0$), the slope with the exponential root has a 20% higher FOS ratio at the slope surface than that with the parabolic root. After raining for 1 h, there was almost no difference in the FOS ratio between the two root architectures. Although the FOS ratio of the vegetated slopes decreased dramatically within the root zone, there was no change outside the root zone during this period of rainfall. It should be highlighted that the maximum FOS ratio of about 1.15 occurred at a depth of 0.5 m, meaning that the FOS of the vegetated slopes was 15% higher than that of the bare slope. As rainfall continued, the FOS ratio both inside and outside the root zone reduced further, and the beneficial effects of plant transpiration on suction vanished. The calculation results suggest that plant transpiration effects are prominent to slope stability, mainly during the early stage of a rainfall event. Moreover, the root architecture does not play a significant role in slope stabilisation when considering the hydrological effects of plant transpiration.

5.3 ROOT-INDUCED CHANGES IN SOIL HYDRAULIC PROPERTIES (NG ET AL., 2016D, 2018B)

5.3.1 Theoretical modelling

To model the effects of plant roots on soil water retention curves (SWRCs; see recent test results in Section 2.4), the mass–volume relationship and the phase diagram of an unsaturated soil, where part of its air void is occupied by plant roots, are considered (Figure 5.8; Ng et al., 2016d). Accordingly, the void ratio of a rooted soil may be expressed as follows:

$$e = \frac{e_0 - R_v(1 + e_0)}{1 + R_v(1 + e_0)} \tag{5.31}$$

where e_0 is the void ratio of bare soil [–], and R_v is the root volume ratio [mm³/mm³], which is defined as the total volume of roots per unit volume of soil. $R_v = 0$ means that there are

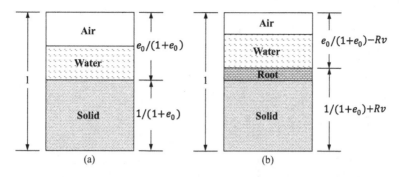

Figure 5.8 Volumetric phase diagram of (a) a bare soil and (b) a rooted soil.

no plant roots in the soil (i.e., the soil is bare). R_v is less than $e_0/(1 + e_0)$, as the total root volume cannot exceed the total soil pore size. Depending on the plant type, R_v is a function of depth within the root zone. Note that it is not the scope of this model to capture any effects of root aging/decay, and the associated formation of macro-pores (Ghestem et al., 2011), on the change in the soil void ratio. Furthermore, for simplicity, the proposed model does not consider any change in the soil micro-structure (i.e., micro-crack development and aggregate formation) during drying–wetting cycles. The proposed model is thus more suitable for low-plasticity materials such as sands and silts.

In order to model the effects of roots on the change in the water retention ability of a soil, the void ratio-dependent SWRC equation proposed by Gallipoli et al. (2003) may be adopted:

$$S_r = \left[1 + \left(\frac{\psi e^{m_4}}{m_3} \right)^{m_2} \right]^{-m_1} \qquad (5.32)$$

where S_r is the degree of soil saturation; ψ is the matric suction; and $m_1[-]$, $m_2[-]$, m_3 [kPa], m_4 [–] are the model parameters. m_1 and m_2 determine the shape of an SWRC (van Genuchten, 1980), while m_3 and m_4 are related to the air-entry value (AEV) of the bare soil. Considering that the void ratio has negligible effects on an SWRC at high suction, the product $m_1 m_2 m_4$ can be set to 1 (Gallipoli et al., 2003). Once the SWRC of the parent soil and the root parameter R_v are known, the SWRC of rooted soil can be predicted.

Figure 5.9 shows a comparison of the measured and predicted SWRC of CDG (silty sand) in the presence and absence of roots of *Schefflera heptaphylla* (see Table 5.1 for the input parameters used). The presence of plant roots increases the soil AEV but does not substantially affect the rate of desorption at relatively large suction. This is consistent with the conceptual model proposed by Scanlan and Hinz (2010), who hypothesise that root occupancy in soil pores reduces the soil pore diameter, causing an increase in suction, following the capillary law.

Correspondingly, the soil water permeability function $k(\psi)$ of a rooted soil may be expressed by van Genuchten (1980) equation as follows:

$$k(\psi) = k_s \cdot S_r^{0.5} [1 - (1 - S_r^{\frac{1}{m_1}})^{m_1}]^2 \qquad (5.33)$$

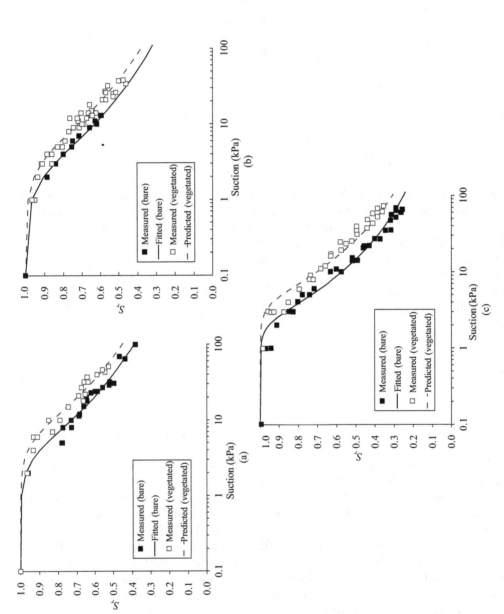

Figure 5.9 Comparison of the measured and predicted SWRCs of bare and vegetated soil from (a) the field tests conducted by Ng et al. (2016d); (b) the laboratory tests conducted by Ng et al. (2016d); and (c) the laboratory tests reported by Leung et al. (2015a).

Table 5.1 Summary of parameters for the SWRC model (Eq. 5.32)

Test	Parameters						
	$m_1 (-)$	$m_2 (-)$	$m_3 (kPa)$	$m_4 (-)$	$e_0 (-)$	Depth (mm)	$R_v (mm^3/mm^3)$
Ng et al. (2016d) (field)	0.11	2.5	0.30	3.64	0.52	100	0.032
Ng et al. (2016d) (laboratory)	0.15	1.9	0.18	3.51	0.50	100	0.034
Leung et al. (2015a) (laboratory)	0.04	8.6	0.70	2.98	0.72	50	0.043

where k_s the saturated water permeability. To capture the fact that k_s could be reduced by the presence of young, intact growing roots (refer to Figures 2.12 and 4.5), the following empirical exponential equation (Yin, 2009) may be used to describe the relationship between k_s and e:

$$k_s = b_1 \cdot \exp(b_2 \cdot e) \tag{5.34}$$

where b_1 and b_2 are the fitting parameters.

5.3.2 Plant effects on matric suction

To highlight the significance of the effects of root-induced soil hydraulic properties on soil hydrology, the *in situ* double-ring infiltration tests conducted by Leung et al. (2015b) are simulated. One can implement the Darcy–Richards equation (Eq. 5.1), together with (1) the SWRC model (Eqs 5.31 and 5.32) and (2) the void ratio-dependent $k(\psi)$ (Eqs 5.33 and 5.34), in MATLAB and solve that equation by using the fully implicit finite-difference method.

To illustrate the effects of roots on matric suction, a field study is selected. According to the site condition, a one-dimensional soil profile with a depth of 4.5 m is modelled. A constant head of 100 mm is specified as the top boundary for 2 h. At the bottom boundary, a constant water table at a depth of 4.5 m is specified. Based on the field data, ψ before ponding is almost completely linearly distributed, and thus, a linear initial distribution of ψ is specified for simulation. More detailed analysis plan and input parameters can be found in Ni et al. (2018b).

Figure 5.10 shows the measured distributions of matric suction (see Table 5.2 for input parameters). Before surface ponding is applied, in the two tests, the initial levels of matric

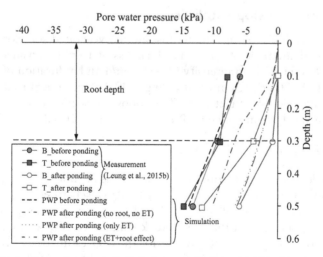

Figure 5.10 Comparison of measured and predicted suction before and after 2 h of ponding from the field study conducted by Leung et al. (2015b).

Table 5.2 Summary of input parameters and boundary conditions for simulating the *in-situ* infiltration tests

| Simulation ID[a] | Input parameters | | | Boundary conditions | |
	Root depth (m)	PE or PT (mm/h)	SWRC and k(ψ)	Top	Bottom
B	N/A	PE: 0.12	Bare soil[b]	Constant water head of 100 mm	Constant groundwater table at a depth of 4.5 m
T	0.3	PT: 0.18	Bare soil[b]		
TR	0.3	PT: 0.18	Tree-covered soil[c]		

[a] 'B' denotes bare soil, 'T' denotes tree-covered soil, 'R' indicates that root-induced changes in soil hydraulic properties are considered in an analysis, PE and PT denote potential evaporation and potential transpiration respectively.

[b] The input soil parameters: $m_1 = 0.11$, $m_2 = 2.5$, $m_3 = 0.30$ kPa, $m_4 = 3.64$, $e_0 = 0.52$ and $k_s = 1.22 \times 10^{-6}$ m/s.

[c] The SWRC and $k(\psi)$ are calculated using Eqs (5.29) through (5.32), with the R_v distributions shown in Ni et al. (2018b) and the parameters summarised in Table 5.1.

Abbreviation: PE: potential evaporation: PT: potential transpiration respectively.

suction in the bare and vegetated grounds are similar. After ponding, however, suction in both the bare and vegetated grounds drops to zero at shallow depths in both tests. At depths below the root zone, the amount of suction preserved in the vegetated ground is always higher than that in the bare ground, by 85%–123%.

The simulation results are superimposed in Figure 5.10 for a direct comparison. When the effects of root-induced changes in soil hydraulic properties are ignored and considering only evapotranspiration (ET), the predicted suction profile of the vegetated ground after ponding is almost identical to that of the bare ground, in both field tests. These small differences are attributed to the small amount of actual transpiration (<0.5 mm) during the 2-h ponding event. The calculated total volume of root water uptake within the root zone is less than 2.8×10^4 mm³, which is 100 times less than the total volume of water infiltrated (>4.3×10^6 mm³), and thus can be neglected. Therefore, root water uptake cannot fully explain the observed suction preserved in the vegetated ground. On the contrary, when root-induced changes in both SWRC and $k(\psi)$ are considered, the predicted suction profiles in both cases are closer to the measurements. This highlights the fact that during a relatively short-lived wetting event, the root-induced changes in soil hydraulic properties are a crucial hydrological effect of vegetation that should not be neglected.

5.3.3 Plant effects on slope stability

Plant effects on the stability of an infinite unsaturated vegetated soil slope are calculated. An infinite slope with an angle of 40° and a thickness of 10 m is considered, and it is subjected to a rainfall event with an intensity of 394 mm/d and a duration of 24 h (equivalent to a return period of 100 years; Lam and Leung, 1995). Mechanical root reinforcement is modelled via root cohesion c_r. Wu et al. (1979) proposed a semi-empirical expression for c_r, and the expression was later modified by Preti and Schwarz (2006):

$$c_r = 0.4 \cdot (\sin \xi_s + \cos \xi_s \tan \phi') \cdot T_r \cdot RAR \tag{5.35}$$

where ξ_s is the angle of shear distortion in the shear zone at root breakage; T_r is the average root tensile strength, which is taken to be 2×10^4 kPa (Bischetti et al., 2005); and RAR is the root area ratio. Wu et al. (1979) showed that the term $(\sin \xi_s + \cos \xi_s \tan \phi')$ is about 1.2. Equation (5.35) assumes that all the roots break simultaneously, without considering any progressive failure of roots. In order to correct for the over-estimation of the root reinforcement due to the assumption made, an empirical correction factor of 0.4 is added by Preti and Schwarz (2006). This factor was calculated as the ratio between the average experimental values and the predicted ones by the model of Wu et al. (1979). Hence, the shear strength of an unsaturated vegetated soil can be determined by the sum of Eqs (5.29) and (5.35).

Hydrological effects of ET are incorporated by including a sink term in Darcy–Richards equation, while the effects of root-induced changes in soil hydraulic properties are considered via Eqs (5.29) through (5.32). By considering both mechanical and hydrological root reinforcements, the FOS of an infinite slope with an inclination of φ at failure can be expressed as (Ni et al., 2018b):

$$\text{FOS} = \frac{\left(c' - u_w \tan\phi^b\right) + 0.48 \cdot T_r \cdot RAR}{\left[\gamma_d\left(H_0 - Z\right) + \gamma_W \int\limits_{Z}^{H_0} \theta dZ\right]\sin\varphi\cos\varphi} + \frac{\tan\phi'}{\tan\varphi} \qquad (5.36)$$

Figure 5.11 shows the FOS of the bare slope and the four vegetated slopes with different profiles of root volume ratio R_v, at the end of the 24-h rainfall event. The top 2.5 m of

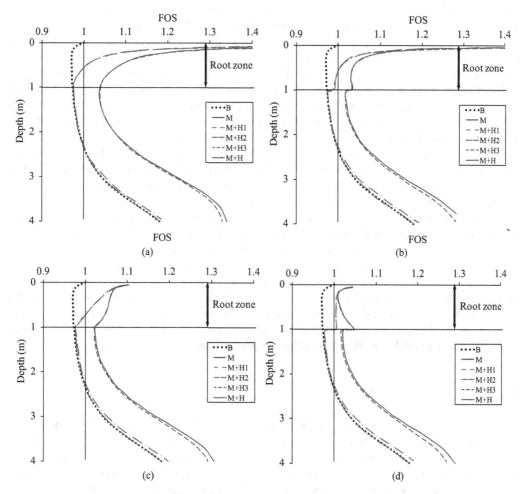

Figure 5.11 Comparison of the effects of hydrological and mechanical reinforcement (lower bound) from vegetation on slope stability after 24 h of rainfall for different profiles: (a) Triangular; (b) uniform; (c) parabolic; and (d) inversely triangular. B is bare soil, and M is mechanical root reinforcement; H1 is the effect of evapotranspiration; H2 is the effect of root-induced changes in SWRCs, and H3 is the effect of root-induced changes in; and H considers all of H1, H2 and H3.

the bare slope was unsafe, as the FOS was less than 1.0. When vegetation with triangular root profile was present (see Figure 5.11a), root mechanical reinforcement only enhanced the slope stability within the root zone. However, the FOS at depths between 0.5 m and 2.5 m was still less than 1.0. The slope could be stabilised only when both mechanical and hydrological effects (i.e., M + H) were considered. Hydrological effects (i.e., a combination of the effects of ET and root-induced changes in soil hydraulic properties; H1 + H2 + H3) contributed additional 7% to the FOS in shallow depths (i.e., top 1 m) and up to 15% in deeper depths below 1 m. The reason of having greater hydrological reinforcement effects in deeper depth is that the suction (hence shear strength) preserved after rainfall is higher below the root zone. Comparison of the four different profiles in Figure 5.11a–d suggests that regardless of the shape of R_v, when all hydrological effects were ignored, only the top 0.5–1 m of the vegetated slope could be stabilised. Neglecting root hydrological reinforcement has significantly underestimated the ability of vegetation for deeper slope stabilisation (up to 2.5 m).

Since ET during the 24-h rainfall event considered in this analysis is minimal (refer to M and M + H1 in Figure 5.11), the removal of soil moisture through root water uptake affects the FOS almost negligibly. Nonetheless, for certain plant species that have high water demand and can transpire during rainy seasons (e.g., *Ulex europaeus* tested in Section 3.2), subjected to long-duration, low-intensity rainfall event, the effects of ET may be more prominent. Indeed, soil column tests conducted by Boldrin et al. (2018) showed that transpiration-induced suction during wetting events was significant for evergreen species, compared with deciduous species, because the former could transpire during rainy seasons, while the latter could not, because of leaf shed. On the other hand, although the presence of roots has shown some effects on SWRCs (see Figures 5.9 and 5.10), this particular hydrological mechanism contributes only little to the slope stability. Predominantly, the hydrological reinforcement is attributed to the root-induced change in k_s. This hydrological effect takes place only within the root zone (refer to Figure 5.11). This, however, has a significant effect on the soil hydrology for the entire soil profile (Leung et al., 2017b).

No major difference in deep hydrological reinforcement is found between the four vegetated cases with different R_v profiles. Consistently, the mechanism of root-induced changes in k_s plays the most significant role in slope stabilisation in all four cases. Arguably, roots with a triangular R_v profile provide a slightly greater hydrological reinforcement below the root zone than the other R_v profiles. The shapes of R_v profiles have a more significant impact on shallow than on deep mechanical reinforcement. While both the triangular and uniform R_v profiles have strong stabilisation effects at very shallow depths (up to 0.5 m), the inversely triangular profile provides relatively weak stabilisation effects (i.e., a smaller increase in the FOS) but has a much larger depth of influence (up to 1 m).

5.4 CHAPTER SUMMARY

Closed-form, analytical steady- and transient-state solutions are derived to capture the effects of four idealised, yet representative, root architectures on the distribution of soil suction and the stability of an unsaturated vegetated soil slope. The calculation shows that vegetated slopes preserve more suction than bare slopes after rainfall, especially at greater depths below the root zone. This is because plant transpiration before rainfall has the effect of drying on the soil, which reduces the unsaturated water permeability. Among the four types of root architectures investigated (i.e., exponential, triangular, uniform and parabolic distributions with depth), the exponential one induces the highest suction and hence is the most effective in stabilising shallow soil slopes.

Further theoretical modelling and analysis reveal that transpiration during intense rainfall is minimal and has a negligible effect on preserving suction and maintaining slope stability (in terms of the FOS) after rainfall. Although plant roots induce a substantial change in the SWRC, they contribute to slope stabilisation only marginally after 100-year return period of rainfall. Two factors play the most significant role in both the magnitude and distribution of matric suction during and after rainfall: (i) soil drying due to plant transpiration before rainfall and (ii) the root-induced change in water permeability.

Chapter 6

Effects of plant on slope hydrology, stability and failure mechanisms

Geotechnical centrifuge modelling

6.1 INTRODUCTION

The previous chapter has clearly demonstrated that plant transpiration and root water uptake can induce and help to preserve a significant amount of matric suction (i.e., negative pore water pressure) during and after rainfall. These kinds of plant hydrological effects can directly increase the soil shear strength and reduce the water infiltration rate. Although a large body of research exists that focuses on the mechanical effects of plant root reinforcement on slope stability, the effects of the identified hydrological effects of plants on slope behaviour are still poorly understood. Moreover, it is difficult (if not impossible) to isolate and quantify the contribution of the mechanical and hydrological effects of vegetation on slope stability under uncontrolled (and undesirable) conditions in the field. It is also extremely costly, not to mention potentially unsafe, to induce failure of real slopes in the field. Since slope stability is a gravity-dependent problem, geotechnical centrifuge modelling is a viable method to correctly study the behaviour and failure mechanisms of slopes under safe, economical and controlled indoor environments. It is a technique that enables scaled physical models to be tested at appropriate stress levels nearly identical to those experienced by much larger prototypes and under much better controlled test conditions than are possible in the field (Taylor, 1995; Ng, 2014).

6.1.1 Fundamental principles of centrifuge modelling (verbatim extract from Ng, 2014)

Soil behaviour is stress dependent. Figure 6.1 illustrates an example, in which a soil sample, A, that is initially located below the critical state line (CSL) will dilate toward the CSL when it is sheared under a relatively low confining stress (e.g., in a small model test under one unit of the Earth's gravity, where $1\ g = 9.81$ m/s^2). By comparison, another sample, B, having the same density (i.e., same void ratio) located at an arbitrary point above the CSL but below or on the normal compression line will contract when it is sheared under a higher mean effective stress, p' (i.e., high stress in the field or in the centrifuge). Obviously using test results from sample A in designing prototype problems is likely to be non-conservative and may even give rise to unsafe situations because the observed dilative behaviour at low stress under $1\ g$ conditions will not occur under high stress in the field. Thus, it is vital to simulate the stress level of the soil correctly before carrying out any physical experiment.

The fundamental principle of centrifuge modelling is to recreate the stress conditions that would exist in a prototype, by increasing the 'gravitational' acceleration N times in a $1/N$-scaled model (where N is the scale factor) in the centrifuge (Taylor, 1995). Stress in the $1/N$-scaled model is approximately replicated by subjecting model components to an elevated 'gravitational' acceleration in the form of centripetal acceleration ($r_c \omega_c^2 = Ng$), where r_c and ω_c

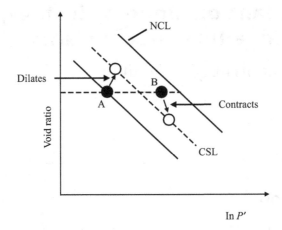

Figure 6.1 Distinct responses of two soil samples at the same density sheared under different confining stresses.

are the radius and angular velocity of the centrifuge, respectively. Thus, a centrifuge can be used to model stress-dependent geotechnical problems. Apart from the ability to replicate in-situ stress in a reduced-size model, one of the side benefits of centrifuge modelling is that the use of a small-scale model shortens the drainage paths of soil, resulting in a significant reduction in consolidation time by $1/N^2$.

For centrifuge model tests, scaling laws are generally derived through dimensional analysis from the governing equations for a phenomenon, or from the principles of mechanical similarity between a model and a prototype (Taylor, 1995; Garnier et al., 2007). Some of the commonly used scaling factors are summarised in Table 6.1.

Table 6.1 Relevant scaling factors for centrifuge tests

Parameter	Scale factor (model/prototype)
Acceleration	N
Linear dimension	$1/N$
Stress	1
Strain	1
Mass	$1/N^3$
Density	1
Unit weight	N
Force	$1/N^2$
Bending moment	$1/N^3$
Bending moment/unit width	$1/N^2$
Flexural stiffness	$1/N^4$
Flexural stiffness/unit width	$1/N^3$
Time (dynamic)	$1/N$
Time (consolidation/diffusion)	$1/N^2$
Time (creep)	1
Pore fluid velocity	N
Velocity (dynamic)	1
Frequency	N

6.1.2 The state-of-the-art geotechnical centrifuge at HKUST (verbatim extract from Ng, 2014)

One of the most advanced geotechnical centrifuges in the world was established at the HKUST in April 2001 (Ng et al., 2001), as shown in Figure 6.2a. This 400 g-t geotechnical centrifuge is equipped with advanced simulation capabilities including the world's first in-flight biaxial (2D) shaker (Shen et al., 1998; Ng et al., 2001; Figure 6.2b), an advanced four-axis robotic manipulator (Ng et al., 2002; Figure 6.2c) and a state-of-the-art data acquisition and control system. This 8.4 m in diameter beam centrifuge is equipped with two swinging platforms, one for static tests and one for dynamic tests. For static tests, the centrifuge is able to accommodate models with dimension of up to 1.5 m × 1.5 m × 1 m. The centripetal acceleration can be up to 150 g. For dynamic tests, the centrifuge incorporates a unique biaxial servo-hydraulic shaker to model earthquake-induced engineering problems (Ng et al., 2001, 2002). The biaxial shaker is capable of simulating earthquake motions in two horizontal directions simultaneously. The shaker can accommodate a mode size of up to 0.6 m × 0.6 m × 0.4 m in size and up to 3000 N in weight. The centrifuge can be operated at up to 75 g for dynamic tests. The advanced and state-of-the-art four-axis robotic manipulator incorporates a tool changer and four tool adopters to permit changing tools in flight. At a centrifugal acceleration of 100 g, the robotic manipulator can produce a torque of up to ±5 MN·m and prototype loads of ±10 MN, ±10 MN, and 50 MN in the x, y, and z directions, respectively.

The HKUST beam centrifuge has been used in the investigation of static liquefaction mechanism of a 30° loose fill slope using Leighton Buzzard (LB) fraction E fine sand subjected to a rising groundwater table (Zhang et al., 2006; Ng, 2007, 2009). Although the initial angle of the loose slope was approximately 30° at 1 g, at 60 g the slope was densified to 80% of the maximum relative compaction under its own weight and thus flattened to 24° (Figure 6.3a), which is less than the critical-state angle of LB sand of 32°. At 60 g, the 18 m high (prototype) slope was destabilised and liquefied by rising groundwater from the bottom of the model (Zhang, 2006). The loose LB sand slope liquefied statically and began flowing rapidly (Figure 6.3b); that is, the loose slope was sheared under undrained conditions, lost its undrained shear strength as a result of the induced high pore-water pressure (PWP) and started flowing like a liquid, called a liquefied flow (Ng, 2007).

6.1.3 Centrifuge modelling of the behaviour of vegetated slopes

This chapter describes the recent novel application of geotechnical centrifuges to the investigation of the hydrological and mechanical effects of vegetation on slope hydrology, stability and failure mechanisms. Real plants of different ages were grown inside a centrifuge to study the mechanical effects of root reinforcement on slope stability. To investigate the combined effects of mechanical reinforcement and plant transpiration-induced matric suction, novel artificial roots with different idealised yet representative architectures were developed to study slope deformations and failure mechanisms at different slope angles (see Chapters 2, 4 and 5 for evidence of their effects). The contributions of different root architectures to slope stability and their failure mechanisms are discussed in detail. Note that all dimensions presented in this chapter are expressed in prototype scale, unless stated otherwise.

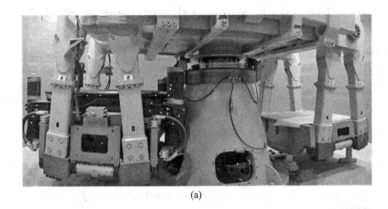

(a)

(b)

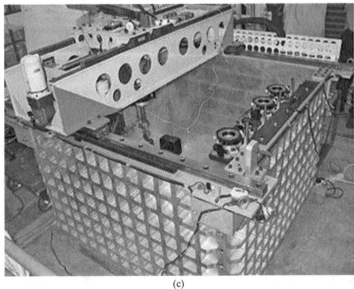

(c)

Figure 6.2 (a) The 8.4 m in diameter (400 g-t) beam centrifuge; (b) bi-axial shaking table; and (c) four-axis robotic manipulator at the HKUST.

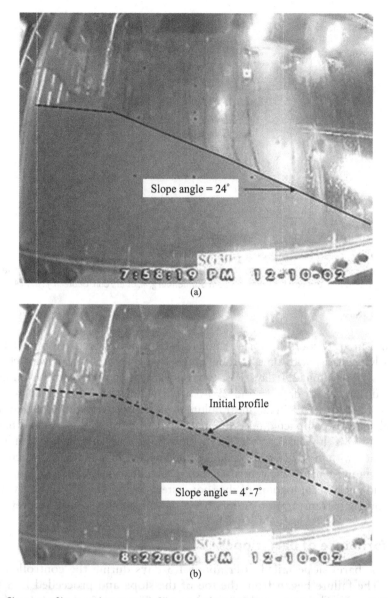

Figure 6.3 (a) Slope profile in a loose sand fill test before rising groundwater table and (b) after static liquefaction. (From Zhang, M. and Ng, C.W.W., Interim Factual Testing Report I-SG30 & SR30, The Hong Kong University of Science and Technology, Hong Kong, China, 2003.)

6.2 MECHANICAL ROOT REINFORCEMENT OF SOIL SLOPES (VERBATIM EXTRACT FROM SONNENBERG ET AL., 2010)

The stability of clayey sand slopes at a scale of 1:15 planted with willow poles 70 mm in length and 4.0 mm in diameter was tested (Sonnenberg et al., 2010). The tensile strength of the willow poles was dependent on the diameter and growth period (Figure 6.4). For any given growth period, the relationship between tensile strength and diameter appeared to follow a negative power law (Eq. [3.1] in Section 3.2). The curves varied greatly. In particular, it would be inappropriate to extrapolate the curves to smaller root diameters

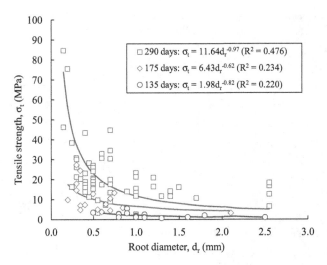

Figure 6.4 Root tensile strength versus root diameter for different growth intervals. σ_t is the root tensile strength, and d_t is the root diameter. (From Sonnenberg, R. et al., *Can. Geotech. J.*, 47, 1415–1430, 2010.)

than were measured (Section 3.3 for a detailed discussion). For a given root diameter, root strength clearly increased with time as roots grew. Two growth intervals were chosen for the centrifuge tests: 29 days and 290 days. Post-test root excavation performed by Sonnenberg et al. (2010) showed that the 29-day-old and 290-day-old willow poles had mean root area ratio (RARs) of 0.035% and 0.031% in the shear failure plane, respectively (Sonnenbereg et al., 2010).

The model soil was compacted to a targeted dry density of 1.24 g/cm³ and had an initial water content of 28% (by mass). In total, 21 willow poles were installed perpendicular to the slope surface at 143 mm × 226 mm (model scale) spacing in a spatially shifted rectangular pattern. The willow stems in the centrifuge box were cut to just above the soil level before testing to avoid mechanical loading of the root systems below the soil surface when the centrifuge was spinning. Transpiration and its effect on soil stability were also prevented.

6.2.1 Performance of bare slopes

The fallow (or bare) slope failed after about 1.3 days during the controlled rise of the water table. The failure began from the toe of the slope and proceeded in a block-wise fashion (Figure 6.5). The first small failure shown in Figure 6.5a occurred with a deforming block that has a maximum depth of 0.48 m. A second failure (Figure 6.5b) was caused by the remaining steep slope section. A third (Figure 6.5c) and a final partial slope failure followed until the overall slope gradient was reduced by these mass movements. Retrieval of the inclinometer installed at the centre of the model after the test revealed a final slope shape that was slightly different from the sides of the model (Figure 6.6). The overall mechanistic behaviour, however, was the same. The measured pore-pressure distribution at failure and the properties of drained soil ($c' = 2.0$ kPa, $\phi' = 24.0°$ and soil unit weight = 16.6 kN/m³; Sonnenberg et al., 2010) were used to calculate a factor of safety (FOS) when failure occurs. The limit equilibrium method suggested by Greenwood (2006) was adopted in the slope stability calculation. The FOS of the bare slope was found to be 0.65 (Sonnenberg et al., 2010), which was consistent with the observation of slope failure in the centrifuge.

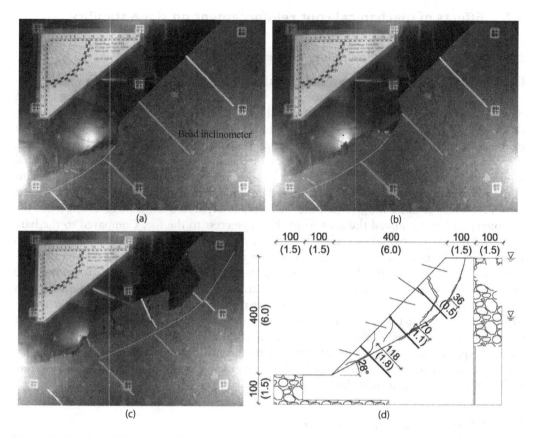

Figure 6.5 Observed failure on a fallow slope (dimensions in millimetres in model scale; prototype scale in metres in parentheses). (a) Digital image of first block failure; (b) second block failure; (c) third block failure; and (d) schematic of the progressive failure mechanism. (From Sonnenberg, R. et al., *Can. Geotech. J.*, 47, 1415–1430, 2010.)

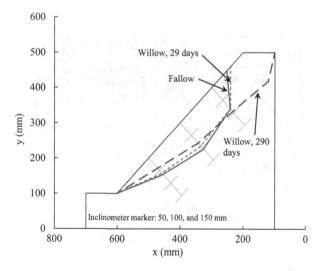

Figure 6.6 Summary of observed failure planes (dimensions in model scale) measured at the box centre. (From Sonnenberg, R. et al., *Can. Geotech. J.*, 47, 1415–1430, 2010.)

6.2.2 Effects of mechanical root reinforcement on slope stability and failure mechanisms

6.2.2.1 Observation of slope failure mode

Like the fallow slope, the slope vegetated with the younger (29-day-old) willow poles displayed a progressive failure mechanism, but with fewer individual blocks. A single large block was observed extending almost from the toe to the crest of the slope. Interestingly, this large failure had a final failure plane depth distribution that was almost identical to that in the fallow test (Figure 6.6) when measured at the box centreline. On the other hand, the slope vegetated with the older willow poles experienced a shallow, mostly translational failure under similar hydraulic conditions to those in the previous tests (Figure 6.7). This figure superimposes the soil displacements determined from digital image analysis using GeoPIV (White et al., 2003). A discrete shear plane separated the sliding block from the stationary soil and spanned the area from the slope crest to the toe. Compared to the bare slope and the 29-day-old willow slope, the 290-day-old willow slope had a deeper failure slip plane, especially near the slope crest. Excavation of root sections (Figure 6.8) revealed significant differences in both below- and above-ground growth between roots grown near the slope toe and those grown close to the slope crest. Due to the model design, willows grown near the toe had less soil to grow in and consequently grew less well than the ones grown in the upper rows, particularly near the crest. The plants in the upper rows featured significant growth of the main structural roots and strong and parallel roots near the soil surface. These roots also grew smaller positively gravitropic lateral roots that were oriented more vertically. The presence of plant roots reinforced the slope, changing the failure mechanism and altering the location of the slip surface during failure.

6.2.2.2 Slope stability back-analysis

In order to quantify the mechanical effects of plant root reinforcement on slope stability, the so-called root cohesion (c_r; defined as a collective term that quantifies the additional increase in soil shear strength) was calculated using the root breakage model proposed by Wu et al. (1979). This model requires information about the RAR and root tensile strength.

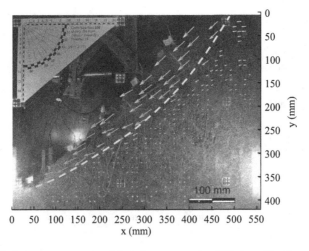

Figure 6.7 Deduced displacement vectors at failure for tests, using 290-day-old willows. (From Sonnenberg, R. et al., *Can. Geotech. J.*, 47, 1415–1430, 2010.)

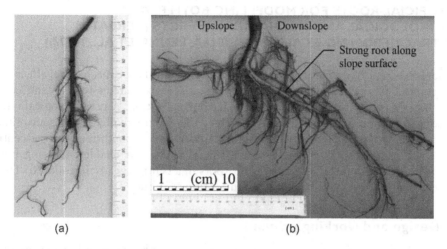

(a)   (b)

Figure 6.8 Excavated roots after slope tests with 290-day-old willows: (a) roots near the toe; and (b) roots near the slope crest. (From Sonnenberg, R. et al., *Can. Geotech. J.*, 47, 1415–1430, 2010.)

The former was obtained by post-test excavation. About 100 roots were excavated from the 29-day-old willow slope and the root diameters generally ranged from 0.7 to 2.6 mm (mean: 1.3 mm; median: 1.2 mm). A mean RAR of 0.035% was revealed at the position of the shear plane. Following the tensile strength-root diameter curve given in Figure 6.4, the average tensile strength for this range of root diameters was 0.10 MPa. Hence, the average c_r of the 29-day-old willow slope was about 0.06 kPa. As for the 290-day-old willow slope, 164 roots were identified above the slip surface and 65 below for one-half of the slope. The average diameter of roots both above and below the slip surface was 0.4 mm, with a median of 0.2 mm. Correspondingly, the RAR varied from 0.018% to 0.048% (mean: 0.031%). Although the RAR in this case was comparable to the values obtained from the younger 29-day-old willows, the mean root tensile strength of 16.1 MPa was two orders of magnitude larger. Thus, based on the relationship proposed by Wu et al. (1979), c_r was 7.7 kPa for the 290-day-old willow slope.

These two calculated values of c_r were subsequently entered into the limit equilibrium calculation proposed by Greenwood (2006). The slope stability back-analysis showed that the slopes with younger and older vegetation at failure had FOS values of 0.57 and 0.7, respectively, consistent with the fact that both vegetated slopes failed in the centrifuge tests. Interestingly, these calculated values of FOS were not too different from that of the fallow slope (0.6). Indeed, based on a visual inspection of the failure plane of both vegetated slopes, no clear distinction could be made between the three root failure models – root breakage, root pull-out and root elongation. It should be emphasised that the values of c_r were estimated using the semi-empirical equation of Wu et al. (1979) based on a bold assumption that all roots fail by breakage and they break simultaneously, which was likely not the case in either centrifuge test. This was one of the major factors leading to the unrealistic estimation of c_r and hence the unrealistic FOS of both vegetated slopes. Other factors that could result in a less accurate calculation of FOS include (i) difficulties in considering the exact root distribution and growth direction in the relatively simple limit equilibrium method; (ii) accuracy of water pressure measurement and hence the identification of the phreatic surface for slope stability analysis; (iii) accuracy of tracing the failure plane and transforming it into a rotational shape in the calculation; and (iv) possible changes in the soil structure/fabric with time during plant growth. A more detailed discussion is given in Sonnenberg et al. (2010).

6.3 ARTIFICIAL ROOTS FOR MODELLING BOTH PLANT HYDROLOGICAL AND MECHANICAL EFFECTS (VERBATIM EXTRACT FROM NG ET AL., 2014B, 2016A; LEUNG ET AL., 2017A)

The model tests reported in Section 6.2 attempted to quantify the mechanical effects of root reinforcement on slope stabilisation. In those tests, the plant stems were intentionally cut to prevent plant transpiration from taking place during testing. Any hydrological effects of plant-induced suction (Chapter 2) were prevented. There are challenges to controlling and quantifying the amount of root water uptake under high-g conditions during centrifuge testing. In this chapter, novel small-scale artificial roots that enable capturing both the hydrological and mechanical effects of plant roots in a centrifuge are introduced. Their use in centrifuge model tests will be described in Sections 6.4 and 6.5.

6.3.1 Design and working principle

Figure 6.9 is a schematic diagram showing the interaction between soil, water, plants, and the atmosphere. Two major physical and biological features (F1 and F2) governing plant transpiration and root-water uptake can be identified as follows:

(F1) During transpiration, water evaporates from plant leaves through the stomata to the atmosphere. Depending on climate conditions, the amount of evaporation is regulated by living guard cells, which control the size of the stomatal opening for the exchange of water vapour with the atmosphere. The decrease in total water head on the surface of plant leaves would create a hydraulic gradient between plant leaves and roots, forcing water to flow upwards (i.e., transpiration pull).

(F2) In response to the hydraulic gradient established by transpiration pull, root-water uptake occurs and the extracted soil moisture is transported upward to plant leaves through the xylem.

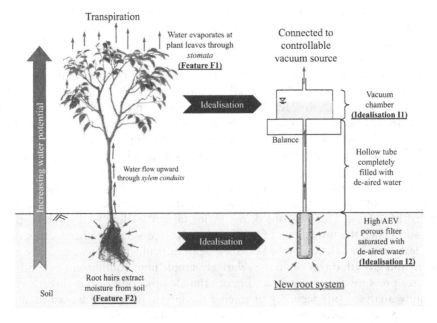

Figure 6.9 Idealisation and simplification of plant transpiration for the development of a novel root system.

The idea of the artificial root system is to idealise and simplify the above two features (F1 and F2). An overview of the artificial root system is shown on the right-hand side of Figure 6.9. This root system consists of two components, a vacuum system (I1) and a porous filter (I2), corresponding to F1 and F2, respectively.

(I1) The vacuum system consists of an air-tight vacuum chamber, which is partially filled with de-aired water. Through a vacuum source connected to the chamber, different vacuum pressures can be applied to induce different pressures of up to −100 kPa in the water reservoir. Regulating the magnitude of water pressure using the vacuum system idealises and simplifies the control of water pressure in leaves by stomata. In the root system, any moisture extracted from the soil due to the applied vacuum can be measured by weighing any mass change in the chamber using an electronic balance.

(I2) The porous filter is connected to the vacuum system through a hollow tube (analogous to a xylem conduit). The porous filter and the tube connecting it to the vacuum system are fully saturated with de-aired water. Since the porous filter is in contact with soil, any applied vacuum, and hence reduction in total head inside the root model, would result in water flowing from the soil to the chamber through the filter (according to Darcy's law). Any decrease in soil moisture would then induce suction in the soil.

Cellulose acetate (CA) was selected as the porous filter material (Figure 6.10). This material is suitable because it has a high air-entry value (AEV) of 100 kPa, so that water pressure

All dimensions are in model scale

Figure 6.10 Overview of a root model made of cellulose acetate.

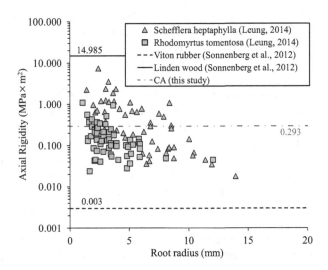

Figure 6.11 Comparison of the axial rigidity of real roots and artificial root models (all units are in metres and in prototype scale). (From Sonnenberg, R. et al., *Can. Geotech. J.*, 49, 1–17, 2012.)

produced by any amount of applied vacuum (up to –95 kPa) in the chamber can be maintained. In addition to simulating the hydrological effect of transpiration-induced suction, CA enables simulating the mechanical root reinforcement. CA has a tensile strength, σ_t, of 31 MPa and an elastic modulus, E_r, of 83 MPa which are fairly close to those typically identified in real roots (Stokes and Mattheck, 1996). As shown in Figure 6.11, the prototype axial rigidity (after scaling up by 15 times; following the scaling factor given in Table 6.2) of CA is close to the average value of real tree roots and other materials that have been used previously for constructing root models. Table 6.2 also summarises the properties of CA in both the model and prototype scales.

Root models with different architectures were developed. Plant roots are assumed to have three typical architectures (Köstler et al., 1968; Stokes and Mattheck, 1996), namely tap, heart and plate (Figure 6.12). The tap-shaped root consists of a primary root and several secondary roots near its tip. The heart-shaped root has a primary taproot, and its secondary roots are normally found near the soil surface and are predominantly sub-horizontal. The plate-shaped root consists of secondary roots fanning out laterally and sub-horizontally at shallow depths. Based on the three idealised root shapes, three simplified root models were constructed. Figure 6.13 shows that the three root models have fairly similar RAR profiles to those of the real roots of *Schefflera heptaphylla*, *Rhodomyrtus tomentosa* and *Melastoma sanguineum*, which are native to Hong Kong.

6.3.2 Performance of the root system

Distribution of suction induced by the tap-shaped artificial root (Test M) in completely decomposed granite (CDG; silty sand) was compared with that of suction induced by a living tree (Test T), *S. heptaphylla*, which is native to Hong Kong, and that of suction induced in bare soil (Test B) (Figure 6.14a). Even after allowing the three specimens to evaporate or evapotranspire in the plant room (refer to Figure 2.1) for 36 h, suction recorded in Test B increased only negligibly because the soil surface was covered with a plastic sheet to minimise surface evaporation. On the contrary, a substantial amount of suction was induced in the top 140 mm (i.e., within the root zone) in Test T due to tree transpiration. Below the root zone, suction increased

Table 6.2 Summary of scaling factors

Physical quantity	Dimension	Scaling factor (model/prototype)	Model scale	Prototype scale[a]
Geometry of artificial roots				
Length	L	$1/N$	50 mm	750 mm
Outer diameter	L	$1/N$	6 mm	90 mm
Inner diameter	L	$1/N$	4 mm	60 mm
Cross-sectional area (A_c)	L^2	$1/N^2$	1.6×10^{-5} m^2	3.5×10^{-3} m^2
Second moment of inertia (I_n)	L^4	$1/N^4$	5.1×10^{-11} m^4	2.6×10^{-6} m^4
Material properties of artificial roots				
Tensile strength of artificial roots (σ_t)	M/LT^{2b}	1	3.1×10^4 kPa	3.1×10^4 kPa
Elastic modulus of artificial roots (E_r)	M/LT^2	1	8.3×10^4 kPa	8.3×10^4 kPa
Axial rigidity (EA) of taproot component	ML/T^2	$1/N^2$	1.3 kPa·m^2	2.9×10^2 kPa·m^2
Flexural rigidity (EI_n) of horizontal root branches	ML^3/T^2	$1/N^4$	4.2×10^{-6} kPa·m^4	2.2×10^{-1} kPa·m^4
Air-entry value of filter	M/LT^2	1	100 kPa	100 kPa
Water permeability of filter	$L/T_{diff}{}^b$	N	2×10^{-6} m/s	1.3×10^{-7} m/s
Soil-atmosphere interface				
Rainfall intensity[c]	L/T_{diff}	N	1,050 mm/h	70 mm/h
Seepage				
Water flow rate[b]	$L^3/T_{diff}{}^b$	$1/N$	Dependent on measurements	
Water permeability[b]	$L/T_{diff}{}^b$	N		
Hydraulic gradient[b]	Unitless	1		
Suction[d]	M/LT^2	1		

[a] Prototype scale at g-level of 15.
[b] Time for dynamic condition (T) is scaled by $1/N$, whereas time for diffusion (T_{diff}) is scaled by $1/N^2$.
[c] According to Taylor (1995).
[d] According to Dell'Avanzi et al. (2004).

slightly from 2 to 5 kPa. The root model in Test M, when subjected to a vacuum pressure of –95 kPa, was capable of inducing a significant amount of suction in the top 140 mm, resulting in a suction profile similar to that induced by the living tree in Test T. The peak suction created by the root model was about 20 kPa, similar to that induced by the plant (i.e., 23 kPa). Horizontal distributions of suction induced by the tree in Test T and the root model in Test M at a depth of 30 mm are compared in Figure 6.14b. For Test T, the amount of suction induced was the largest at 10 mm away from the tree roots, but it decreased with an increase in horizontal distance. A similar horizontal distribution of suction was induced by the artificial root.

The performance of the artificial root system was evaluated in the centrifuge at 15 g. When a constant vacuum pressure of 65 kPa was consistently applied (Test C-1; Figure 6.15a), PWP recorded within the root zone (i.e., the top 1.5 m) showed substantial decreases from positive to negative values. After three days of water uptake by the root models, the peak suction induced at a depth of 0.45 m was up to 17 kPa. At the greater depth of 2.2 m, some reduction in PWP was observed, but the decrease was not as significant as that found at shallower depths. When a higher constant vacuum pressure of –98 kPa was applied (Test C-2), the suction profile was similar to that obtained in Test C-1, but the magnitudes of suction

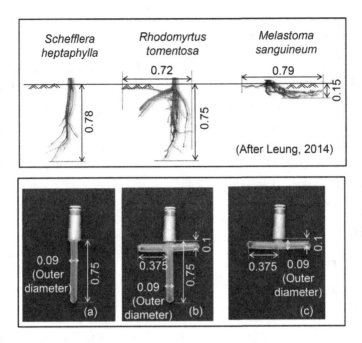

Figure 6.12 Overview of (a) tap-, (b) heart- and (c) plate-shaped root models (all dimensions are in metres and in prototype scale).

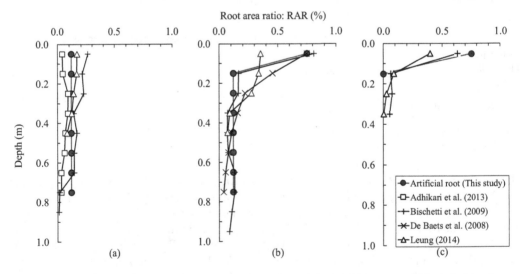

Figure 6.13 Comparison of the root area ratio (RAR) profiles of (a) tap-, (b) heart- and (c) plate-shaped root models in prototype scale and those of real roots observed in the field. (From Adhikari, A.R. et al., *Ecol. Eng.*, 51, 33–44, 2013.)

induced at both the depths of 0.45 and 1.2 m were higher. The peak suction induced at the depth of 0.45 m at the end of the test reached 25 kPa.

Horizontal distributions of induced suction in Tests C-1 and C-2 are compared in Figure 6.15b. When a constant vacuum pressure was applied in each test, the amount of suction was the highest closest to the central root model but decreased with an increase in horizontal distance. However, as the horizontal distance increased further from 0.75 to 1.2 m,

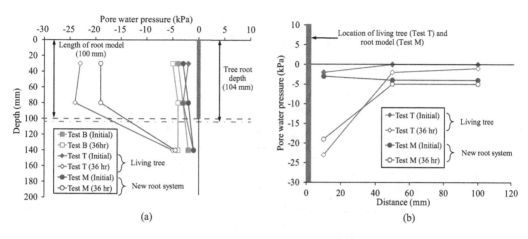

Figure 6.14 Comparison of measured (a) vertical and (b) horizontal distributions of pore water pressure induced by the root system (Test M) and a living tree (Test T) at 1 g.

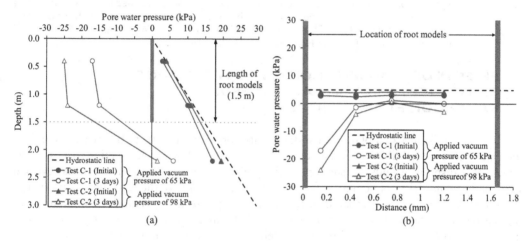

Figure 6.15 Comparison of the (a) vertical and (b) horizontal distributions of pore water pressure induced in Tests C-1 and C-2 conducted at 15 g in the centrifuge.

suction increased, which was not observed in Test M in the laboratory (see Figure 6.14). Such increase in suction was attributed to the water uptake action created by the root model 'vegetated' 1.7 m away from the central one. The observed distributions of suction in both Tests C-1 and C-2 were the consequence of the overlapping influence zones of suction induced by the two neighbouring root models.

The centrifuge-measured PWP profiles induced by the root model in Test C-2 are compared with the living tree measurements made in Test T at 1 *g* and the results of Test M at 1 *g* in Figure 6.16. In addition, field-measured PWP profiles from relevant case histories including test results obtained from the HKUST Eco-Park (see Figure 4.12; Section 4.2) are also included in the comparison. The PWP profiles obtained from Tests T and M at 1 *g* were consistent with each other. Furthermore, suction induced by the root model in the centrifuge was fairly consistent across the case histories, given that the ground conditions and the environment in the field were not completely identical to those simulated in the centrifuge. Despite observed discrepancies between the centrifuge data and field observations, it was evident that the root model could create suction that fell within a reasonable range retained by vegetation in the field.

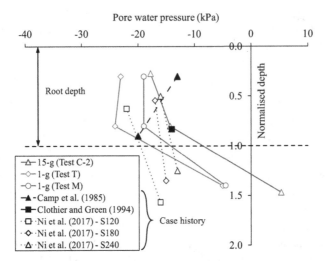

Figure 6.16 Comparison of the distributions of induced pore water pressure obtained from 1 *g* Test M, 15 *g* Test C-2 and various case histories after a 36-h drying period. (From Camp, C.R. et al., *Trans ASAE*, 28(4), 1159–1165, 1985; Clothier, B.E. and Green, S.R., *Agric. Water Manag.*, 25, 1–2, 1994; Myers, J.M. et al., *Proc. Fla. State Hort. Soc.*, 89, 23–28, 1976.)

6.4 EFFECTS OF TRANSPIRATION ON ROOT PULL-OUT RESISTANCE (VERBATIM EXTRACT FROM KAMCHOOM ET AL., 2014)

The pull-out resistance of a plant root system contributes importantly to the mechanical reinforcement of slopes (Waldron and Dakessian, 1981; Ennos, 1990). This section discusses the effects of transpiration-induced suction on the pull-out resistance of plant roots as revealed via centrifuge model tests. Three different architectures, namely the tap-, heart- and plate-shaped architectures, are compared in terms of their pull-out behaviour. To simulate both the mechanical effects of root reinforcement and the hydrological effects of transpiration-induced suction, the novel artificial root models developed in Section 6.3 were used.

6.4.1 Effects of root architecture on the PWP distribution (i.e., matric suction)

The model box (Figure 6.17) contained five artificial roots spaced (1.73 m) evenly in a uniform layer of compacted CDG (clayey sand at RC of 95%; see Kamchoom et al. (2014) for its properties). An identical vacuum pressure was applied to all artificial roots to simulate the effects of transpiration for five days. Figure 6.18 shows that during this process, suction (i.e., negative PWP) within the depths of both tap- and heart-shaped roots (Figure 6.18a and b) increased substantially. Because the heart-shaped root had two branches, it was able to induce a higher suction. For the plate-shaped roots which did not have any taproot component, suction was noticeable only at a depth of 0.3 m (Figure 6.18c). After the simulation of transpiration, rainfall with a constant intensity of 70 mm/h was applied for 2 h (equivalent to a return period of 10 years). Suction at all depths in all three types of roots was reduced as expected (Figure 6.19). After rainfall, similar suction ranging from 5 to 10 kPa was preserved within the depths of both the tap- and heart-shaped roots (Figure 6.19a and b). As rainfall continued for another 6 h with the same intensity (equivalent to a return period of 1000 years), positive PWP of about 2 kPa resulted within the root depth. However, the plate-shaped roots showed

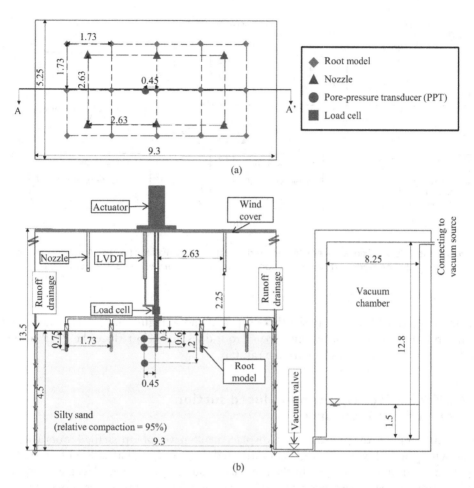

Figure 6.17 (a) Top view and (b) elevation view of section A–A′ of a centrifuge model package and instrumentation for pull-out tests of the tap-shaped artificial roots (all dimensions in metres and in prototype scale). (From Kamchoom, V. et al., *Geotech. Lett.*, 4, 330–336, 2014.)

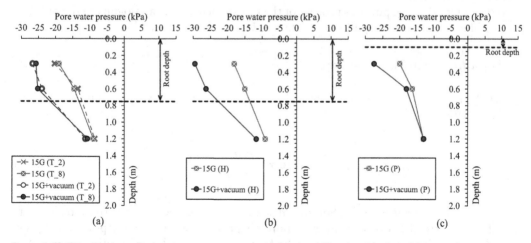

Figure 6.18 Distribution of pore water pressure with depth at 15 g for (a) tap-, (b) heart- and (c) plate-shaped root models. (From Kamchoom, V. et al., *Geotech. Lett.*, 4, 330–336, 2014.)

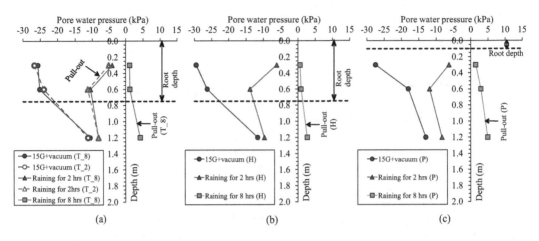

Figure 6.19 Distribution of pore water pressure with depth after rainfall for (a) tap-, (b) heart- and (c) plate-shaped root models. (From Kamchoom, V. et al., *Geotech. Lett.*, 4, 330–336, 2014.)

the highest positive PWP among the three types of roots (Figure 6.19c). This suggests that any transpiration carried out by this type of root does not help reduce the PWP below the root depth very effectively under extreme rainfall.

6.4.2 Effects of transpiration-induced suction on pull-out resistance

Figure 6.20 compares the pull-out force-displacement curves of tap-shaped roots after 2 h of rainfall when the soil was still largely unsaturated (Test T_2) and after 8 h of rainfall when the soil was saturated (Test T_8). In the presence of suction, (i) the displacement required to fully mobilise the peak resistance and (ii) the peak pull-out resistance were both higher than the case without suction. After reaching the peak (4.74 kN for T_2 and 3.54 kN for T_8), the pull-out force in both cases dropped as the soil-root contact area reduced continuously during the pull-out process.

To the extent that the pull-out process of the tap-shaped root is analogous to the pull-out process of an axial pile, the pull-out resistance, F_p, may be described by:

$$F_P = \pi d_r L \overline{\sigma_h}' \tan(\delta), \qquad \text{where } \overline{\sigma_h}' = K \overline{\sigma_v}'. \tag{6.1}$$

$\overline{\sigma_v}'$ may be expressed by an instance of Bishop's equation that was suggested by Khalili and Khabbaz (1998):

$$\overline{\sigma_v}' = \left(\overline{\sigma_v} - \overline{u_a}\right) + \chi\left(\overline{u_a} - \overline{u_w}\right), \qquad \text{where } \chi = \left[\frac{\left(\overline{u_a} - \overline{u_w}\right)}{AEV}\right]^{-\xi}. \tag{6.2}$$

where $\overline{\sigma_h}'$ is the average effective horizontal stress, $\overline{\sigma_v}$ is average vertical stress, $\overline{\sigma_v}'$ is average effective vertical stress, $\overline{u_a}$ is average pore air pressure, $\overline{u_w}$ is average pore water pressure, d_r is root diameter, L is root length, K is coefficient of lateral earth pressure, δ is interface friction angle determined from direct shear tests, χ is Bishop's parameter and

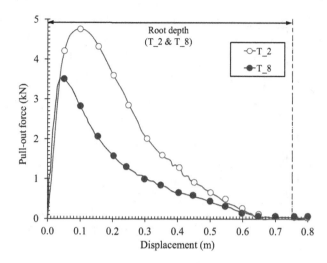

Figure 6.20 Comparison of the pull-out behaviour of tap-shaped root models measured after 2 (Test T_2) and 8 (Test T_8) hours of rainfall. (From Kamchoom, V. et al., *Geotech. Lett.*, 4, 330–336, 2014.)

ξ is the material parameter (Kamchoom et al., 2014). Based on the PWP measurements (Figure 6.19) and the peak pull-out resistance, the back-calculated K for both Tests T_2 and T_8 was 3.3, which was far from K_0 (i.e., 0.398, as estimated by Jaky's equation) but close to K_p (i.e., 4.02 using Rankine's theory) of CDG. The relatively large K values suggest that considerable constraint dilatancy (Houlsby, 1991) had likely taken place upon intense shearing at the soil-root interface, causing $\overline{\sigma_h}'$ to be significantly higher than $\overline{\sigma_v}'$. In the presence of suction (Test T_2), soil dilatancy increased (Chiu and Ng, 2003). This led to a higher $\overline{\sigma_h}'$ and explained the observed higher peak pull-out resistance than that in the saturated case (Test T_8).

6.4.3 Effects of root architecture on pull-out resistance

Additional centrifuge tests were conducted to measure the pull-out behaviour of the other two root architectures, namely the heart-shaped (Test H) and plate-shaped (Test P) architectures. Both pull-out tests were carried out after 2 h of rainfall. Figure 6.21 shows that the peak resistance of the heart-shaped root (3.9 kN) was not much higher than that of the tap-shaped root (3.5 kN; Test T_8). Moreover, the post-peak behaviour of the former architecture was more brittle, as less pull-out displacement was required to mobilise the same amount of resistance in the former than in the latter. Given the similar positive PWP induced after rainfall (Figure 6.19a and b), the difference in pull-out behaviour between these two architectures was primarily attributed to the mechanical reinforcement given by the two branches of the heart-shaped root model. However, their contribution to pull-out appeared marginal because the difference in pull-out resistance was small. The horizontal branches were ineffective at mobilising soil-root interface friction against the vertical pull-out. Although the root branches were bent and provided some flexural resistance, the degree of mobilisation was considered minimal as (i) the E_r of the root material was higher than that of the soil (see Table 6.2) and (ii) the branches were located at depths where soil overburden pressure (i.e., <2 kPa) was not significant.

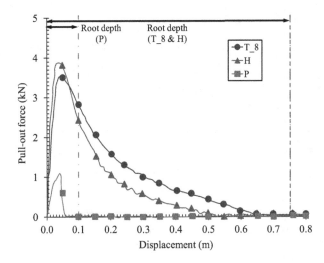

Figure 6.21 Comparison of the pull-out behaviour of tap-, heart- and plate-shaped root models measured after 8 h of rainfall. (From Kamchoom, V. et al., *Geotech. Lett.*, 4, 330–336, 2014.)

Unlike the previous two cases, the plate-shaped root did not appear to be effective against pull-out because (i) the peak pull-out resistance (1.2 kPa) was three to four times lower and (ii) the post-peak load-displacement response is much more brittle. The major reason was that for this root architecture, the two horizontal branches located at much shallower depths were not able to provide much pull-out resistance. The observation was in agreement with the comparison of field pull-out tests made by Nilaweera and Nutalaya (1999). They found that the peak pull-out force of *Hibiscus macrophyllus* (whose root system tends to be more heart-shaped) was about 4 kN, which was 50% higher than that of *Alstonia macrophylla*, a plant that has a more plate-shaped root system.

6.5 PLANT HYDRO-MECHANICAL EFFECTS ON SLOPE BEHAVIOUR (VERBATIM EXTRACT FROM NG ET AL., 2016A; LEUNG ET AL., 2017A)

This section discusses the effects of different plant root architectures on the hydrology, stability and failure mechanisms of unsaturated vegetated soil slopes subjected to intense rainfall. Three 15 *g* centrifuge tests were conducted using the geotechnical centrifuge at HKUST, considering three different root architectures – tap, heart and plate (Figure 6.12). In each test, a 45° compacted slope made of CDG (silty sand; Ng et al., 2016a) was 'vegetated' with a total of 15 artificial roots arranged in three columns by five rows with a spacing of 1.73 m (Figure 6.22). After reaching 15 *g* in each test, all artificial roots were supplied with the same vacuum pressure of 95 kPa to simulate the effects of plant transpiration. After suction equilibrium, each model slope was subjected to an intense rainfall event with a constant intensity of 70 mm/h for 2 and 8 h (equivalent to a return period of 10 and 1000 years, respectively, according to the rainfall data recorded in Hong Kong; Lam and Leung, 1995).

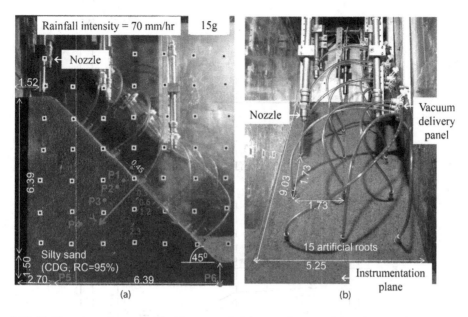

Figure 6.22 (a) Elevation view and (b) side view of the centrifuge model package and instrumentation (all dimensions in metres and in prototype scale).

6.5.1 Effects of plant root architecture on slope hydrology

6.5.1.1 Responses of pore-water pressure during the simulation of transpiration

Figures 6.23a–c show the measured PWP variations with time of slopes containing tap-, heart- and plate-shaped roots, respectively. Prior to spinning up of the centrifuge, the initial PWP recorded by all pore pressure transducers (PPTs) was about –15 kPa for all three slope models. As the g-level rose, PWP jumped at all instrument locations in all three slope models due to self-weight consolidation. When 15 g was reached, the PWP fell, but much more slowly than it rose when g-level was increasing. This was attributed to the dissipation of the excess PWP generated previously. When a vacuum pressure of –95 kPa was applied to simulate the effects of transpiration, PWP fell substantially at the depths of 0.3 and 0.6 m in the slopes supported by the tap- and heart-shaped roots (Figures 6.23a and b). On the contrary, for the slope supported by the shallower plate-shaped roots (Figure 6.23c), PWP showed a reduction only at the depth of 0.3 m. Root architecture thus plays a significant role in the distribution of induced suction. When PWP reached equilibrium after the application of vacuum, rainfall was applied. Upon infiltration, the PPTs in the top 1.2 m of soil in all three slopes showed significant increases in PWP to around –1 and –2 kPa and further to a positive value of about 10 kPa at a depth of 2.3 m after 8 h of rainfall. As the rainfall event continued, the measured increases in PWP were less significant, and they appeared to have reached the steady state.

Figure 6.24 compares measured PWP responses with depth among the three slopes. At 15 g, the measured PWP profiles in all three slopes were very similar. When the vacuum pressure of 95 kPa was applied, the measured PWP within the depth of the

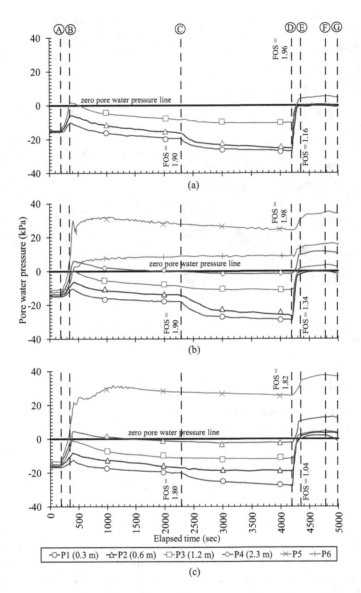

Figure 6.23 Measured variations in pore water pressure with time for slope models supported by (a) tap-, (b) heart- and (c) plate-shaped root models.

tap-shaped root (Figure 6.24a) decreased substantially from a value of about –20 to –26 kPa. On the contrary, PWP below the root depth of 1.2 m showed only a slight decrease of about 2 kPa. PWP at a depth of 0.3 m was 15% higher in the slope supported by heart-shaped roots (Figure 6.24b) than in the one supported by tap-shaped roots (Figure 6.24a). This was because the heart-shaped roots had two additional root branches (Figure 6.24a), which served to induce a greater reduction in PWP. However, since the two branches were located at a relatively shallow depth of 0.1 m, they were only able to exert an influence on PWP up to a depth of 0.3 m. This explained why the PWP induced at a depth of 0.6 m was similar between the two cases (i.e., about

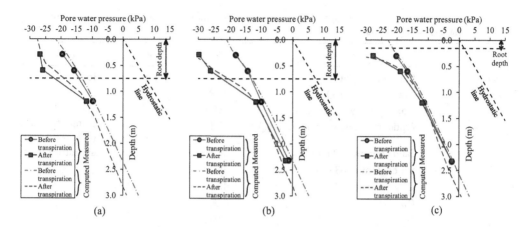

Figure 6.24 Distribution of measured and computed pore water pressure with depth before and after 'transpiration' for slope models supported by (a) tap-, (b) heart- and (c) plate-shaped root models.

–26 kPa). For the slope supported by plate-shaped roots (Figure 6.24c), induced PWP appeared to be affected mainly at the shallow depths where the two branches were (i.e., at 0.3 m), but it was much less significantly affected at greater depths.

To assist the interpretation of the centrifuge-observed soil hydrological responses and to perform subsequent slope stability analysis, finite element transient seepage back-analysis was carried out using SEEP/W (Geo-Slope Int., 2009). The CA material that made up the artificial roots was modelled as a kind of porous material, whose air-entry value was set to 100 kPa and whose saturated water permeability was set to 1×10^{-6} m/s according to Ng et al. (2014b). Other input parameters are summarised in Table 6.3. To simulate the effects of transpiration adopted by the artificial roots, a constant pressure head was applied along the internal boundary of these model roots. It can be seen from Figure 6.24 that in general, the computed PWP profiles agreed well with the measurements made in all three cases, both before and after transpiration. It is thus safe to say that the simulation of transpiration in the centrifuge was fairly accurately back-analysed when a constant head was applied along the root boundaries.

6.5.1.2 Responses of pore-water pressure during rainfall

Figure 6.25 compares the variations in the water infiltration rate with time among the three slopes. As the rainfall was maintained at a fixed intensity of 70 mm/h, the initial infiltration rates in the first hour remained constant in all three slopes. However, when the infiltration capacity was reached, all rates fell exponentially, approaching the saturated water permeability conductivity of the soil. The infiltration rate in the slope supported by the heart-shaped roots was lower than that in the slope supported by the tap-shaped roots, while the one supported by the plate-shaped roots had the highest infiltration rate. The observed trend was consistent with the measured PWP responses. As shown in Figure 6.24, the PWP induced in the case of heart-shaped roots at the shallowest measurement depth of 0.3 m was the lowest, which would have caused the greatest reduction in water permeability among the three cases. Although similar amounts of PWP were induced by tap- and plate-shaped roots (Figure 6.24), the former induced a much lower PWP at a depth of 0.6 m than

Table 6.3 Summary of soil and root properties and the input parameters used in the finite element seepage-stability analysis

	Parameter		Value	Unit	References
Soil index properties	Bulk unit weight (γ_t)		20	kN/m³	Ho (2007)
	Specific gravity (G_s)		2.59	–	
	Maximum dry density (ρ_{max})		1890	kg/m³	Standard Proctor compaction tests (BSI, 1990)
	Optimum moisture content (w_{opt})		15.1	%	
	Sand content (≤2 mm)		56.8		Laboratory tests
	Silt content (≤63 μm)		39.1		
	Clay content (≤2 μm)		4.1		
	D_{10}		0.005	mm	
	D_{30}		0.041		
	D_{50}		0.081		
	D_{60}		0.115		
	Plastic limit (*PL*)		22.7	%	
	Liquid limit (*LL*)		32.8		
	Plasticity index (*PI*)		10.1		
Mechanical properties of soil	Effective cohesion (c')		0	kPa	Hossain and Yin (2010)
	Critical-state friction angle (ϕ_{cr}')		37.4	degree	
	Dilation angle (ϕ_d)		5	degree	
	Young's modulus (E)		35	MPa	Zhou (2008)
	Poisson ratio (μ)		0.26	–	
Hydraulic properties of soil	Saturated water permeability (k_s)		1×10^{-7}	m/s	Falling head tests
	Air-entry value (*AEV*)		1	kPa	Ho (2007)
	Saturated water content (θ_s)		41	%	SWRCs measured by Ho (2007) and then fitted by van Genuchten (1980)
	Residual water content (θ_r)		15		
	Fitting parameters for the van Genuchten (1980) equation	a	0.5	kPa⁻¹	
		n	1.8	–	
		m	0.44		
Mechanical properties of artificial roots	Tensile strength (σ_t)		3.1×10^4	kPa	Root tensile tests (Ng et al., 2014b)
	Young's modulus (E)		8.3×10^4	kPa	
	Interface friction angle (δ)		34	degree	Direct shear tests (Kamchoom et al., 2014)
Hydraulic properties of artificial roots	Saturated water permeability (k_s)		2×10^{-6}	m/s	Ng et al. (2014b)
	Air-entry value (*AEV*)		100	kPa	

the latter. This led to the difference between their water infiltration rates. For the purpose of seepage modelling, the curve of each measured infiltration rate was best-fitted. Each fitted line was then superimposed on the slope surface boundary in SEEP/W to simulate the wetting event (Figure 6.25).

Figure 6.26 compares the measured PWP profiles of the three slopes during rainfall infiltration. For the slope supported by tap-shaped roots (Figure 6.26a), PWP showed the most significant increase at the shallowest depth of 0.3 m after 2 h of rainfall (equivalent to a

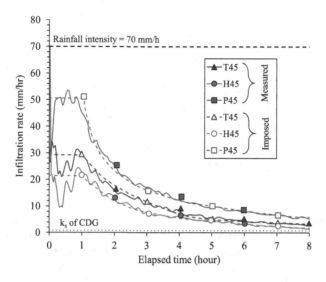

Figure 6.25 Measured variations in the infiltration rate with time and the best-fitted lines used in seepage analysis. k_s denotes saturated water permeability.

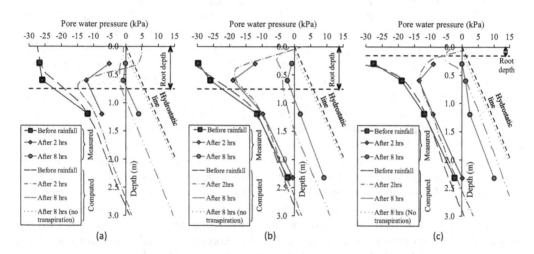

Figure 6.26 Distribution of measured and computed pore water pressure with depth before rainfall and after 2 and 8 h of rainfall in slope models supported by (a) tap-, (b) heart- and (c) plate-shaped roots.

10-year return period), whereas the increase was smaller at depths below the root zone. The seepage analysis also revealed similar suction at the three instrument depths, although a positive PWP of about 5 kPa was measured within 0.3 m of the slope surface. As rainfall continued for another 6 h (equivalent to a 1000-year return period), PWP at all depths increased further. However, the amount of PWP increase within the root depth was much smaller than that below the root depth. As a result, suction of 2 kPa was preserved at the shallower depths, while a positive PWP of about 4 kPa was built up at a depth of 1.2 m. This highlights the significance of the effects of transpiration on suction redistribution. Figure 6.27a shows the computed PWP contour after 8 h of rainfall for the case of the tap-shaped roots. A suction zone was preserved around the roots, especially those roots located

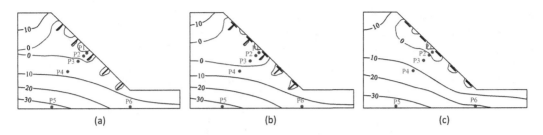

Figure 6.27 Computed pore water pressure contours after 8 h of rainfall in slopes supported by (a) tap-, (b) heart- and (c) plate-shaped root models.

in the upper part of the slope. Suction was preserved because transpiration reduced the soil water permeability and hence the infiltration rate at the later stage of the rainfall event. Below the root depth where PWP was less affected by transpiration, the PWP profile exhibited an almost hydrostatic distribution (Figure 6.26a) and a groundwater table (GWT) was developed (Figure 6.27a). The seepage analysis gave a reasonably accurate prediction of the PWP recorded in the slope.

For the slope supported by heart-shaped roots (Figure 6.26b), the responses of PWP were largely similar to those observed in the case of tap-shaped roots, even though in the former case (i) the initial PWP before rainfall was considerably lower (by about 15%); and (ii) the infiltration rate (Figure 6.25) was slightly lower. After raining for 8 h, measurements showed that similar levels of suction (of 2–3 kPa) were retained within the root zone in the cases of tap- and heart-shaped roots. However, the computed PWP contour depicted in Figure 6.27b shows that the suction zone developed in the case of heart-shaped roots was larger than that found in the case of tap-shaped roots (Figure 6.27a). This suggests that the additional root branches of heart-shaped roots helped extend the influence zone of transpiration to affect the PWP regime of the slope in deeper regions. A similar GWT was found in both cases after 8 h of rainfall.

The PWP responses for the case of plate-shaped roots, which had a much shallower depth (Figure 6.26c), measured during the first 2 h of rainfall were again similar to those for the cases of tap- and heart-shaped roots. However, as rainfall continued, very little suction was retained in the root zone, while positive PWP values were recorded at all depths and they were distributed hydrostatically. These measured PWP responses were consistent with the computed PWP contour shown in Figure 6.27c. The contour showed that mainly localised suction was retained near the roots and the suction zone developed around the roots was smaller than those found in the cases of tap- and heart-shaped roots. This was mainly because the plate-shaped roots did not have a taproot component, and thus their induced suction had less effects at greater depths. Although the computed PWP at the end of the rainfall event was slightly underestimated in the top 0.5 m of soil in the case of plate-shaped roots (Figure 6.27c), PWP values at greater depths were fairly close to the measurements. Given the limitation of seepage analysis and the simplification made, the discrepancies between measured and computed PWP are considered acceptable.

6.5.2 Plant effects on slope stability

Slope stability analysis was performed using SIGMA/W (Geo-Slope Int., 2009). In each analysis, the PWP distributions computed from SEEP/W were entered into SIGMA/W for stress-deformation analysis and FOS determination using the strength reduction method

(SRM; Dawson et al., 1999; Griffiths and Lane, 1999). The principle of SRM is to apply a factor that continuously reduces the shear strength parameters (i.e., effective cohesion c' and effective friction angle ϕ' for the Mohr-Coulomb model, for instance) until the slope can no longer maintain its equilibrium. During the shear strength reduction, imbalanced force would be developed in the slope. The strength reduction factor that causes a non-equilibrium condition of the slope is referred to as the FOS. More details about the principle of the SRM can be found in Griffiths and Lane (1999) and Geo-Slope Int. (2009). In each stability analysis, the CDG was considered as a perfectly-plastic material obeying the modified extended Mohr-Coulomb failure criterion proposed by Vanapalli et al. (1996) (Eq. 1.2). On the other hand, the artificial roots were modelled as a beam element (Hinton and Owen, 1979) to capture both the elastic axial and bending responses. The input parameters associated with the soil and the artificial roots are reported in detail in Ng et al. (2014b, 2016a).

Figure 6.28 compares the FOS of the slopes supported by the three different root architectures. Before transpiration, the FOS was similar for all three slopes and exceeded 1.0, meaning that all slopes were stable as consistently observed in all centrifuge tests. Figure 6.29a–c show the corresponding mobilisation of shear strains in the slope before transpiration for tap-, heart- and plate-shaped roots, respectively. The initial suction at shallow depths, together with the mechanical reinforcement attributed to the root models, provided sufficient strength for the shallow soil to remain stable. Shear strain thus occurred at greater depths below the root zone. Localised shear strain of up to 200% was mobilised, indicating the potential formation of a slip plane where the sliding mass above would fail. As strain mobilisation mainly occurred at depths below the root zone, the stability of all slopes at depths of up to 1 m was adequately maintained by the roots. When suction was created by the simulation of transpiration, the FOS of each slope increased, although not by much (less than 4%). The corresponding shear strain contours in Figure 6.29d–f revealed that strain localisation clearly occurred at the same location before transpiration. This was because the applied transpiration mainly affected the PWP in the top 1.2 m of each slope (see Figure 6.24). This suggests that induced suction provides additional stabilisation effects at relatively shallow depths (i.e., within 1.2 m of the soil surface).

After raining for 8 h, the FOS of all three slopes dropped significantly (Figure 6.28), following the reduction in PWP upon infiltration. Shear strain intensified substantially near the toe of each slope after the rainfall (Figure 6.29g–i). Despite the reduction in the FOS,

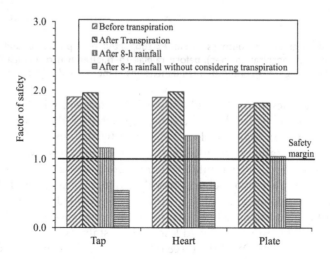

Figure 6.28 The FOS of slopes supported by different root architectures.

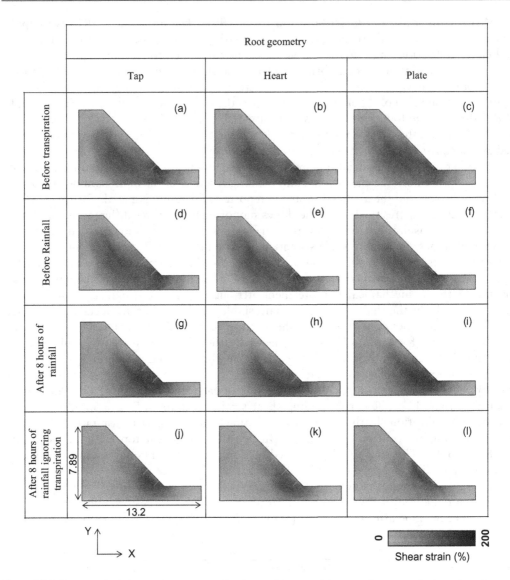

Figure 6.29 Computed shear strain contours in slopes supported by tap-, heart- and plate-shaped root models before transpiration (a–c), before rainfall (d–f), after 8 h of rainfall (g–i) and after 8 h of rainfall ignoring transpiration (j–l) (all dimensions in metres and in prototype scale).

the values in all three cases were still higher than 1.0, consistent with the centrifuge observation that all slopes remained stable. The slope supported by the heart-shaped roots had a 16% and 28% higher FOS than the slope supported by the tap- and plate-shaped roots, respectively. The greater stability provided by the heart-shaped roots was mainly due to the substantial suction preserved after rainfall (Figure 6.26) and their higher mechanical pull-out resistance than the other two architectures (Figure 6.21). As none of the plate-shaped roots had a taproot component, the pull-out resistance provided by this root architecture was only one-third of that mobilised in the tap- and heart-shaped roots (Figure 6.21), and therefore this root architecture contributed less to slope stability.

Figure 6.28 also shows the FOS of the three types of slopes ignoring the effects of transpiration before rainfall. The FOS values after rainfall were all less than 1.0, indicating potential slope failure. The major reason was that when the effects of transpiration were ignored, no suction was preserved in all three slopes and a significant amount of positive PWP was developed hydrostatically in the soil after the rainfall (see Figure 6.26). Following such a substantial increase in PWP, strain localisation occurred in the shallower region and closer to the toe of each slope (Figure 6.29j–l). Regardless of the root architecture, neglecting the effects of transpiration on slope stability resulted in a significant drop in the FOS by up to 50%.

6.5.3 Effects of root architecture on the failure mechanisms of 60° slopes

In order to further explore the effects of plant root architecture on slope failure mechanisms, two additional centrifuge tests were conducted for much steeper, 60° slopes 'vegetated' with tap- and heart-shaped artificial roots (Tests T60 and H60, respectively). Similar test procedures to those adopted in testing the 45° slopes were used, i.e. applying an identical constant vacuum pressure to all artificial roots followed by an intense rainfall event with a constant intensity of 70 mm/h for 8 h (equivalent to a rainfall return period of 1000 years in Hong Kong). At the end of the rainfall event, both 60° slopes failed.

Figure 6.30a and b depict the measured PWP responses upon rainfall infiltration in these two additional tests. In general, the PWPs within the root zone (i.e., the top 0.6 m) in both slopes showed rapid increases during the first one and a half hours of rainfall, whereas the PWPs at the greater depth of 1.2 m rose more gradually. In Test T60, shallow slope failure occurred after 3 h of rainfall when the PPT at the depth of 0.3 m recorded suction of 3 kPa (Figure 6.30a). The PPT then lost contact with the surrounding soil. Further shallow slope

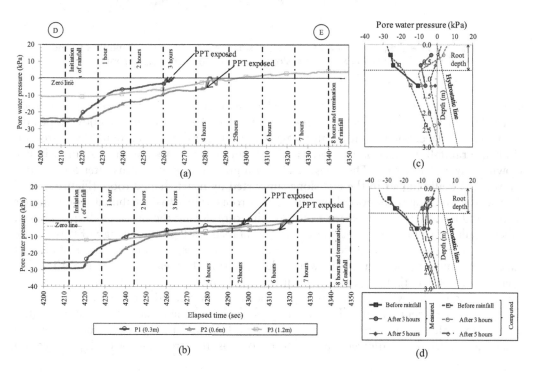

Figure 6.30 Comparison of the variations in suction with time for (a) Test T60 and (b) Test H60, and the pore water pressure profiles for (c) Test T60 and (d) Test H60 during rainfall.

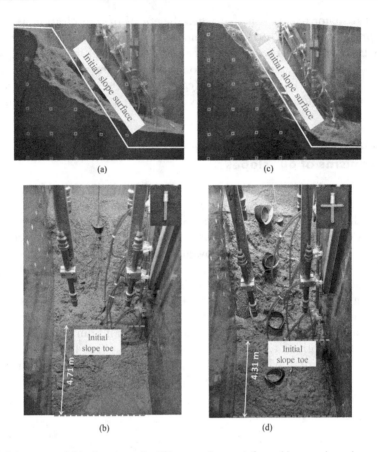

Figure 6.31 (a) Side view and (b) plan view of a 60° steep slope reinforced by tap-shaped roots, and (c) side view and (d) plan view of a 60° steep slope reinforced by heart-shaped roots.

failure occurred after nearly 5 h of rainfall when the suction at the depth of 0.6 m reached 6 kPa. This PPT also lost contact with the soil subsequently. In Test H60, shallow slope failure also took place, but at a much later stage, after 5 h of rainfall (Figure 6.30b). After 8 h of intense rainfall, no suction was left in all tests, but both slopes remained stable without showing any global failure. The measured suction preserved at all depths after 5 h of rainfall was markedly higher in the 60° slope supported by heart-shaped roots (Figure 6.30d) than in the 60° slopes supported by tap-shaped roots. This was attributed to the higher suction induced by the heart-shaped root architecture during the previous drying period, which resulted in a lower infiltration rate. The computed PWP profiles using SEEP/W before and after rainfall are also depicted in Figure 6.30c and d for comparison. The simulations matched the measurements and they were able to capture the measured drop in PWP.

Figure 6.31 shows that slope failure occurred at relatively shallower depths and a smaller volume of soil failed when the heart-shaped root model was used for slope reinforcement. The heart-shaped root reinforced slope remained stable for the longest duration of rainfall for 5 h (equivalent to a return period of 250 years; Lam and Leung, 1995), beyond which some slips occurred at shallower depths. The runout distance from the toe of slope vegetated with tap-shaped roots was about 10% longer than that of slope reinforced with heart-shaped roots. These observations suggest the importance of considering root architecture for slope stability. Heart-shaped roots appear to be more effective at stabilising slopes and reducing landslide hazards than tap-shaped roots.

6.6 CHAPTER SUMMARY

Stability and failure mechanisms of vegetated sandy slopes during rainfall and rising GWT were investigated using centrifuge modelling techniques. The presence of vegetation changed the slope failure mechanisms and modes. By considering the mechanical effects of roots only, progressive block-wise failure was observed for bare slopes, whereas planted slopes exhibited intact sliding block failure with slightly deeper slips than bare slopes. Back analysis of the observed slope failure showed that the FOS values of both bare and vegetated slopes were less than 1.0. However, there was no discernible difference in the FOS, no matter whether the mechanical effects of plant roots (the so-called root cohesion) were considered or not. This was probably because the assumption made in estimating the root cohesion and hence the slope stability (i.e., all roots cutting through a slip fail by breakage simultaneously) did not accord well with the conditions found in the vegetated slopes tested in the centrifuge.

After developing novel artificial model root systems, it is now possible to simulate not only the mechanical effects of root reinforcement but also the hydrological effects of transpiration-induced suction on the hydrology, stability and failure mechanisms of vegetated soil slopes in a geotechnical centrifuge at the correct stress level. Artificial roots of three different idealised yet representative architectures, namely the tap, heart and plate architectures, have been created. The development of these novel artificial root systems creates new opportunities for studying geotechnical and geoenvironmental problems such as vegetated landfill covers and other engineering problems involving vegetation.

Regardless of the root architecture, none of the 45° slopes failed after experiencing a 1000-year rainfall event, owing to contributions from both transpiration-induced suction and mechanical root reinforcement. Stability analysis also revealed that root architecture played a significant role in slope stability. The heart-shaped roots provided more stabilisation effects to the slope than either the tap- or plate-shaped roots. This was because the heart-shaped roots preserved more suction after rainfall (owing to their larger soil–root hydraulic contact area for water uptake) than the other two architectures. Hence, this reduced the infiltration rate during rainfall. In addition, the heart-shaped roots provided higher pull-out resistance. For all three root architectures studied, ignoring the soil shear strength attributed to transpiration-induced suction resulted in much lower FOS values (more than 50% lower than that in the case with transpiration simulation).

All of the steeper 60° slopes, which were supported by the same two root architectures (tap- and heart-shaped), failed after the same 1000-year rainfall event. Although all slopes failed, the one supported by the heart-shaped roots remained stable for the longest duration of rainfall for 5 h, beyond which some slips occurred at shallower depths. Moreover, a smaller volume of soil failed in the slope supported by the heart-shaped roots, while the runout distance from the slope toe was also the shortest. This suggests that the heart-shaped roots are more effective in stabilising steep slopes and reducing landslide hazards than the tap-shaped roots.

References

Adhikari, A. R., Gautam, M. R., Yu, Z., Imada, S. and Acharya, K. (2013). Estimation of root cohesion for desert shrub species in the Lower Colorado riparian ecosystem and its potential for streambank stabilization. *Ecological Engineering*, 51, 33–44.

AFCD. (2010). *Care for Our Trees*. The Government of the Hong Kong Special Administrative Region, Hong Kong, China.

AFCD. (2012). Characteristics of major local tree species propagated by AFCD. http://www.afcd.gov. hk/english/country/cou_lea/plantation_tree.html. Accessed 10 December 2013.

Aitchison, G. D. (1965). *Moisture Equilibria and Moisture Changes in Soils Beneath Covered Areas; A Symposium in Print*. Butterworths, Sydney, Australia, 278 pp.

Alam, J., Das, A., Rahman, M. and Islam, M. (2015). Effect of waterlogged condition on wood properties of *Acacia nilotica* (L.) Debile tree. *Bangladesh Journal of Scientific and Industrial Research*, 50(2), 71–76.

Allen, R. G., Pereira, L. S., Raes, D. and Smith, M. (1998). *Crop Evapotranspiration: Guidelines for Computing Crop Water Requirements*. FAO Irrigation and Drainage Paper 56. FAO, Rome, Italy.

Andrews, M., Raven, J. A. and Lea, P. J. (2013). Do plants need nitrate? The mechanisms by which nitrogen form affects plants. *Annals of Applied Biology*, 163, 174–199.

Arnold, D. H. and Mauseth, J. D. (1999). Effects of environmental factors on development of wood. *American Journal of Botany*, 86(3), 367–371.

ASTM. (2009). *Standard Test Method for Infiltration Rate of Soils in Field Using Double-Ring Infiltrometer*. American Society for Testing and Materials, West Conshohocken, PA.

ASTM. (2011). *Standard Practice for Classification of Soils for Engineering Purposes (Unified Soil Classification System)*. American Society for Testing and Materials, West Conshohocken, PA.

Azam-Ali, S. N., Gregory, P. J. and Monteith, J. L. (1984). Effects of planting density on water use and productivity of pearl millet (*Pennisetum typhoides*) grown on stored water: Growth of roots and shoots. *Experimental Agriculture*, 20(3), 203–214.

Bai, K., He, C., Wan, X. and Jiang, D. (2015). Leaf economics of evergreen and deciduous tree species along an elevational gradient in a subtropical mountain. *AoB Plants*, 7. doi:10.1093/aobpla/plv064.

Balestrini, R. and Bonfante, P. (2005). The interface compartment in arbuscular mycorrhizae: A special type of plant cell wall? *Plant Biosystems*, 139, 8–15.

Balestrini, R. and Bonfante, P. (2014). Cell wall remodeling in mycorrhizal symbiosis: A way towards biotrophism. *Frontiers in Plant Science*, 5, 1–10.

Balestrini, R., Romera, C., Puigdomenech, P. and Bonfante, P. (1994). Location of a cell-wall hydroxyproline-rich glycoprotein, cellulose and β-1,3-glucans in apical and differentiated regions of maize mycorrhizal roots. *Planta*, 195, 201–209.

Barker, D. H. (1995). Vegetation and slopes: Stabilisation, protection and ecology. In *Proceedings of the International Conference*, University Museum, Thomas Telford, Oxford, UK, 29–30 September 1994.

Bashline, L., Li, S. and Gu, Y. (2014). The trafficking of the cellulose synthase complex in higher plants. *Annals of Botany*, 114(6), 1059–1067.

Bedini, S., Pellegrino, E., Avio, L., Pellegrini, S., Bazzoffi, P., Argese, E. and Giovannetti, M. (2009). Changes in soil aggregation and glomalin-related soil protein content as affected by the arbuscular mycorrhizal fungal species *Glomus mosseae* and *Glomus intraradices*. *Soil Biology and Biochemistry*, 41, 1491–1496.

Beikircher, B., Florineth, F. and Mayr, S. (2010). Restoration of rocky slopes based on planted gabions and use of drought preconditioned woody species. *Ecological Engineering*, 36(4), 421–426.

Bengough, A. G. (2012). Water dynamics of the root zone: Rhizosphere biophysics and its control on soil hydrology. *Vadose Zone Journal*, 11(2). doi:10.2136/vzj 2011.0111.

Bengough, A. G. and Mullins, C. E. (1990). Mechanical impedance to root growth: A review of experimental techniques and root growth responses. *Journal of Soil Science*, 41(3), 341–358.

Bischetti, G. B., Chiaradia, E. A., Epis, T. and Morlotti, E. (2009). Root cohesion of forest species in the Italian Alps. *Plant and Soil*, 324, 71–89.

Bischetti, G. B., Chiaradia, E. A., Simonato, T., Speziali, B., Vitali, B., Vullo, P. and Zocco, A. (2005). Root strength and root area ratio of forest species in Lombardy (northern Italy). *Plant and Soil*, 278, 11–22.

Blight, G. E. (1997). Interactions between the atmosphere and the earth. *Géotechnique*, 47(4), 715–767.

Bloom, A. J., Meyerhoff, P. A., Taylor, A. R. and Rost, T. L. (2002). Root development and absorption of ammonium and nitrate from the rhizosphere. *Journal of Plant Growth Regulation*, 21(4), 416–431.

Bochet, E. and García-Fayos, P. (2015). Identifying plant traits: A key aspect for species selection in restoration of eroded roadsides in semiarid environments. *Ecological Engineering*, 83, 444–451.

Bodner, G., Scholl, P. and Kaul, H.-P. (2013). Field quantification of wetting–drying cycles to predict temporal changes of soil pore size distribution. *Soil and Tillage Research*, 133, 1–9.

Boldrin, D., Leung, A. K. and Bengough, A. G. (2017a). Root biomechanical properties during establishment of woody perennials. *Ecological Engineering*, 109, 196–206.

Boldrin, D., Leung, A. K. and Bengough, A. G. (2017b). Correlating hydromechanical properties of vegetated soil with plant functional traits. *Plant and Soil*. doi:10.1007/s11104-017-3211-3.

Boldrin, D., Leung, A. K. and Bengough, A. G. (2018). Hydrologic reinforcement induced by contrasting woody species during summer and winter. *Plant and Soil*, 427(1–2), 369–390.

Bonfante, P. and Genre, A. (2010). Mechanisms underlying beneficial plant–fungus interactions in mycorrhizal symbiosis. *Nature Communications*, 1, 48.

Bonfante, P., Vian, B., Perotto, S., Faccio, A. and Knox, J. P. (1990). Cellulose and pectin localization in roots of mycorrhizal *Allium porrum*: Labelling continuity between host cell wall and interfacial material. *Planta*, 180(4), 537–547.

Bouriaud, O., Leban, J.-M., Bert, D. and Deleuze, C. (2005). Intra-annual variations in climate influence growth and wood density of Norway spruce. *Tree Physiology*, 25(6), 651–660.

Briggs, K. M., Smethurst, J. A., Powrie, W. and O'Brien, A. S. (2016). The influence of tree root water uptake on the long term hydrology of a clay fill embankment. *Transportation Geotechnics*, 9, 31–48.

Brodersen, C. R., McElrone, A. J., Choat, B., Matthews, M. A. and Shackel, K. A. (2010). The dynamics of embolism repair in xylem: In vivo visualizations using high-resolution computed tomography. *Plant Physiology*, 154(3), 1088–1095.

BSI. (1990). *Methods of Test for Soils for Civil Engineering Purposes*; standard BS 1377–4:1990. British Standards Institution, London, UK.

BSI. (2010). *BSI British Standards on Tree Work: Recommendations (BS 3998:2010)*. British Standards Institution, London, UK.

Buchanan, B. B., Gruissem, W. and Russell, L. J. (2009). *Biochemistry & Molecular Biology of Plants*. American Society of Plant Physiologists, Derwood, MD.

Burylo, M., Hudek, C. and Rey, F. (2011). Soil reinforcement by roots of six dominant species on eroded mountainous marly slopes (Southern Alps, France). *Catena*, 84, 70–78.

Caird, M. A., Richards, J. H. and Donovan, L. A. (2007). Nighttime stomatal conductance and transpiration in C3 and C4 plants. *Plant Physiology*, 143(1), 4–10.

Camp, C. R., Karlen, D. L. and Lambert, J. R. (1985). Irrigation scheduling and row configurations for corn in the southeastern coastal plain. *Transactions of the ASAE*, 28(4), 1159–1165.

Campbell Scientific Inc. (2008). *Instruction Manual: LI190SB Quantum Sensor*. Campbell Scientific Inc., Logan, UT.

Carminati, A., Zarebanadkouki, M., Kroener, E., Ahmed, M. A. and Holz, M. (2016). Biophysical rhizosphere processes affecting root water uptake. *Annals of Botany*, 118, 561–571.

Casaroli, D., de Jong van Lier, Q. and Neto, D. D. (2010). Validation of a root water uptake model to estimate transpiration constraints. *Agricultural Water Management*, 97, 1382–1388.

Casper, B. B. and Jackson, R. B. (1997). Plant competition underground. *Annual Review of Ecology and Systematics*, 28, 545–570.

Chen, X. W., Kang, Y., So, P. S., Ng, C. W. W. and Wong, M. H. (2018). Arbuscular mycorrhizal fungi increase the proportion of cellulose and hemicellulose in the root stele of vetiver grass. *Plant and Soil*. doi:10.1007/s11104-018-3583-z.

Cheung, K. Y. (2007). *Landscape Plants*. Morning Star, Taipei, Taiwan.

Chiatante, D., Sarnataro, M., Fusco, S., Di Iorio, A. and Scippa, G. S. (2003). Modification of root morphological parameters and root architecture in seedlings of *Fraxinusornus* L. and *Spartium junceum* L. growing on slopes. *Plant Biosystems*, 137(1), 47–55.

Chiu, C. F. (2001). Behaviour of unsaturated loosely compacted weathered materials, PhD thesis, Hong Kong University of Science and Technology, Hong Kong, China.

Chiu, C. F. and Ng, C. W. W. (2003). A state-dependent elastoplastic model for saturated and unsaturated soils. *Géotechnique*, 53(9), 809–829.

Clothier, B. E. and Green, S. R. (1994). Rootzone processes and the efficient use of irrigation water. *Agricultural Water Management*, 25, 1–2.

Coppin, N. J. and Richards, I. G. (1990). *Use of Vegetation in Civil Engineering*. Construction Industry Research and Information Association/Butterworths, London, UK, 292 pp.

Corlett, R. T. (1998). Frugivory and seed dispersal by birds in Hong Kong shrubland. *Forktail*, 13, 23–27.

Corlett, R. T. (2001). Pollination in a degraded tropical landscape: A Hong Kong case study. *Journal of Tropical Ecology*, 17, 155–161.

Corlett, R. T., Xing, F. W., Ng, S. C., Chau, L. K. C. and Wong, L. M. Y. (2000). Hong Kong vascular plants: Distribution and status. *Memoirs of the Hong Kong Natural History Society*, 23, 1–157.

Cosgrove, D. J. (2014). Re-constructing our models of cellulose and primary cell wall assembly. *Current Opinion in Plant Biology*, 22, 122–131.

Cross, A. F. and Schlesinger, W. H. (1995). A literature review and evaluation of the Hedley fractionation: Applications to the biogeochemical cycle of soil phosphorus in natural ecosystems. *Geoderma*, 64(3–4), 197–214.

Cutler, D. F., Botha, T. and Stevenson, D. W. (2009). *Plant Anatomy: An Applied Approach*. Wiley-Blackwell, Oxford, UK.

Danjon, F., Barker, D. H., Drexhage, M. and Stokes, A. (2008). Using three-dimensional plant root architecture in models of shallow-slope stability. *Annals of Botany*, 101(8), 1281–1293.

Danjon, F., Khuder, H. and Stokes, A. (2013). Deep phenotyping of coarse root architecture in R. pseudoacacia reveals that tree root system plasticity is confined within its architectural model. *PLoS One*, 8(12), e83548.

Darawsheh, M. K., Khah, E. M., Aivalakis, G., Chachalis, D. and Sallaku, F. (2009). Cotton row spacing and plant density cropping systems I: Effects on accumulation and partitioning of dry mass and LAI. *Journal of Food, Agriculture and Environment*, 7(3–4), 258–261.

Dawson, E. M., Roth, W. H. and Drescher, A. (1999). Slope stability analysis by strength reduction, *Geotechnique*, 49(6), 835–840.

Dawson, T. E., Burgess, S. S. O., Tu, K. P., Oliveira, R. S., Santiago, L. S., Fisher, J. B., Simonin, K. A. and Ambrose, A. R. (2007). Nighttime transpiration in woody plants from contrasting ecosystems. *Tree Physiology*, 27(4), 561–575.

De Baets, S., Poesen, J., Reubens, B., Wemans, K., De Baerdemaeker, J. and Muys, B. (2008). Root tensile strength and root distribution of typical Mediterranean plant species and their contribution to soil shear strength. *Plant and Soil*, 305, 207–226.

de Boer, H. J., Lammertsma, E. I., Wagner-Cremer, F., Dilcher, D. L., Wassen, M. J. and Dekker, S. C. (2011). Climate forcing due to optimization of maximal leaf conductance in subtropical vegetation under rising CO_2. *Proceedings of the National Academy of Sciences*, 108(10), 4041–4046.

DeJong, J. T., Burbank, M., Kavazanjian, E., Weaver, T., Montoya, B. M., Hamdan, N., Bang, S. S. et al. (2013). Biogeochemical processes and geotechnical applications: Progress, opportunities and challenges. *Geotechnique*, 63(4), 287–301.

DeJong, J., Tibbett, M. and Fourie, A. (2015). Geotechnical systems that evolve with ecological processes. *Environmental Earth Science*, 73, 1067–1082.

Dell'Avanzi, E., Zornberg, J. G. and Cabral, A. (2004). Suction profiles and scale factors for unsaturated flow under increased gravitational field. *Soils and Foundations*, 44(3), 1–11.

Dong, X. W., Gen, X. W., Yong, F. B., Jian, X. L. and Hong, X. R. (2002). Response of growth and water use efficiency of spring wheat to whole season CO_2 enrichment and drought. *Acta Botanica Sinica*, 44, 1477–1483.

Dorioz, J. M., Robert, M. and Chenu, C. (1993). The role of roots, fungi and bacteria on clay particle organization: An experimental approach. *Geoderma*, 56(1–4), 179–194.

DSD (Drainage Services Department). (2013). *Stormwater Drainage Manual*. Hong Kong Government, Hong Kong, China.

Dumlao, M. R., Ramananarivo, S., Goyal, V., DeJong, J. T., Waller, J. and Silk, W. K. (2015). The role of root development of Avena fatua in conferring soil strength. *American Journal of Botany*, 102(7), 1050–1060.

Edlefsen, N. E. and Anderson, A. B. (1943). Thermodynamics of soil moisture. *Hilgardia*, 15, 31–298.

Eissenstat, D. M. (1992). Costs and benefits of constructing roots of small diameter. *Journal of Plant Nutrient*, 15(6–7), 763–782.

Ennos, A. R. (1990). The anchorage of leek seedlings: The effect of root length and soil strength. *Annals of Botany*, 65(4), 409–416.

Fan, C. C. and Su, C. F. (2008). Role of roots in the shear strength of root-reinforced soils with high moisture content. *Ecological Engineering*, 33, 157–166.

Fatahi, B., Khabbaz, H. and Indraratna, B. (2010). Bioengineering ground improvement considering root water uptake model. *Ecological Engineering*, 36(2), 222–229.

Feddes, R. A., Kowalik, P., Kolinska-Malinka, K. and Zaradny, H. (1976). Simulation of field water uptake by plants using a soil water dependent root extraction function. *Journal of Hydrology*, 31(1), 13–26.

Feddes, R. A., Kowalik, P. J. and Zaradny, H. (1978). *Simulation of Field Water Use and Crop Yield*. John Wiley & Sons, New York.

Fiorilli, V., Catoni, M., Miozzi, L., Novero, M., Accotto, G. P. and Lanfranco, L. (2009). Global and cell-type gene expression profiles in tomato plants colonized by an arbuscular mycorrhizal fungus. *New Phytologist*, 184, 975–987.

Fontaine, S., Barot, S., Barré, P., Bdioui, N., Mary, B. and Rumpel, C. (2007). Stability of organic carbon in deep soil layers controlled by fresh carbon supply. *Nature*, 450(7167), 277–280.

Francour, P. and Semroud, R. (1992). Calculation for the root area index in *Posidonia oceanica* in the Western Mediterranean. *Aquatic Botany*, 42(3), 281–286.

Fredlund, D. G. and Rahardjo, H. (1993). *Soil Mechanics for Unsaturated Soils*. Wiley, New York.

Fredlund, D. G., Xing, A. and Huang, S. (1994). Predicting the permeability functions for unsaturated soils using the soil-water characteristic curve. *Canadian Geotechnical Journal*, 31, 533–546.

Gabr, M. A., Akran, M. and Taylor, H. M. (1995). Effect of simulated roots on the permeability of silty soil. *Geotechnical Testing Journal*, 18(1), 112–115.

Gallipoli, D., Wheeler, S. J. and Karstunen, M. (2003). Modelling the variation of degree of saturation in a deformable unsaturated soil. *Géotechnique*, 53(1), 105–112.

Galloway, J. N., Townsend, A. R., Erisman, J. W., Bekunda, M., Cai, Z., Freney, J. R., Martinelli, L. A., Seitzinger, S. P. and Sutton, M. A. (2008). Transformation of the nitrogen cycle: Recent trends, questions and potential solutions. *Science*, 320(5878), 889–892.

Gan, J. K. M., Fredlund, D. G. and Rahardjo, H. (1988). Determination of the shear strength parameters of an unsaturated soil using the direct shear test. *Canadian Geotechnical Journal*, 25(3), 500–510.

Gardner, W. R. (1958). Some steady-state solutions of the unsaturated moisture flow equation with application to evaporation from a water table. *Soil Science*, 85(4), 228–232.

Gardner, W. R. (1960). Dynamic aspects of water availability to plants. *Soil Science*, 89(2), 63–73.

Garg, A., Coo, J. L. and Ng, C. W. W. (2015a). Field study on influence of root characteristics on suction distributions in slopes vegetated with *Cynodon dactylon* and *Schefflera heptaphylla*. *Earth Surface Processes and Landforms*, 40(12), 1631–1643.

Garg, A., Leung, A. K. and Ng, C. W. W. (2015b). Comparisons of soil suction induced by evapotranspiration and transpiration of *S. heptaphylla*. *Canadian Geotechnical Journal*, 52(12), 2149–2155.

Garnier, J., Gaudin, C., Springman, S. M., Culligan, P. J., Goodings, D., Konig, D., Kuttervii, B. et al. (2007). Catalogue of scaling laws and similitude questions in geotechnical centrifuge modelling. *International Journal of Physical Modelling in Geotechnics*, 7(3), 1–23.

Gates, D. M. (1980). *Biophysical Ecology*. Springer-Verlag, New York.

Genet, M., Stokes, A., Fourcaud, T. and Norris, J. E. (2010). The influence of plant diversity on slope stability in a moist evergreen deciduous forest. *Ecological Engineering*, 36, 265–275.

Genet, M., Stokes, A., Salin, F., Mickovski, S. B., Fourcaud, T., Dumail, J.-F. and Beek, R. (2005). The influence of cellulose content on tensile strength in tree roots. *Plant and Soil*, 278, 1–9.

GEO (Geotechnical Engineering Office). (2011). *Technical Guidelines on Landscape Treatment for Slopes*. Geotechnical Engineering Office, Hong Kong, China.

Geo-Slope International Ltd. (2009). *Seepage Modelling with SEEP/W, An Engineering Methodology*, 4th ed. Geo-Slope International Ltd., Calgary, Canada.

Ghestem, M., Cao, K., Ma, W., Rowe, N., Leclerc, R., Gadenne, C. and Stokes, A. (2014). A framework for identifying plant species to be used as 'Ecological Engineers' for fixing soil on unstable slopes. *PLoS One*, 9. doi:10.1371/journal.pone.0095876.

Ghestem, M., Sidle, R. C. and Stokes, A. (2011). The influence of plant root systems on subsurface flow: Implications for slope stability. *Bioscience*, 61(11), 869–879.

Giovannetti, M. and Mosse, B. (1980). An evaluation of techniques for measuring vesicular arbuscular mycorrhizal infection in roots. *New Phytologist*, 84, 489–500.

Gish, T. J. and Jury, W. A. (1983). Effect of plant roots and root channels on solute transport. *Transactions of the American Society of Agricultural and Biological Engineers*, 26(2), 440–444.

Granovsky, A. V. and McCoy, E. L. (1997). Air flow measurements to describe field variation in porosity and permeability of soil macropores. *Soil Science Society of America Journal*, 61(6), 1569–1576.

Gray, D. H. and Barker, D. (2004). Root-soil mechanics and interactions. In Bennett, J. J. and Simon, A. (Eds.), *Riparian Vegetation and Fluvial Geomorphology*, Water Science and Applications, vol. 8. American Geophysical Union, Washington, DC, pp. 113–123.

Gray, D. H. and Leiser, A. T. (1982). *Biotechnical Slope Protection and Erosion Control*. Van Nostrand Reinhold Company Inc., New York.

Gray, D. H. and Sotir, R. B. (1996). *Biotechnical and Soil Bioengineering Slope Stabilization: A Practical Guide for Erosion Control*. Wiley, New York.

Greenwood, J. R. (2006). SLIP4EX: A program for routine slope stability analysis to include the effects of vegetation, reinforcement and hydrological changes. *Geotechnical and Geological Engineering*, 24(3), 449–465.

Greenwood, J. R., Norris, J. E. and Wint, J. (2004). Assessing the contribution of vegetation to slope stability. *Proceedings of the Institution of Civil Engineers–Geotechnical Engineering*, 157(4), 199–207.

Gregory, P. J. (2008). *Plant Roots: Growth, Activity and Interactions with the Soil*. Wiley-Blackwell, Oxford, UK.

Griffiths, D. V. and Lane, P. A. (1999). Slope stability analysis by finite elements. *Geotechnique*, 49(3), 387–403.

Grime, J. P., Thompson, K., Hunt, R., Hodgson, J. G., Cornelissen, J. H. C., Rorison, I. H., Hendry, G. A. F. et al. (1997). Integrated screening validates primary axes of specialisation in plants. *Oikos*, 79, 259–281.

Grouzis, M. and Akpo, L.-E. (1997). Influence of tree cover on herbaceous above- and below-ground phytomass in the Sahelian zone of Senegal. *Journal of Arid Environments*, 35(2), 285–296.

Hamblin, A. P. and Tennant, D. (1987). Root length density and water uptake in cereals and grain legumes: How well are they correlated. *Crop and Pasture Science*, 38(3), 513–527.

Harris, R. W., Clark, J. R. and Matheny, N. P. (2004). *Arboriculture: Integrate Management of Landscape Trees, Shrubs, and Vines*. Prentice Hall, Upper Saddle River, NJ.

Hau, B. C. H. and Corlett, R. T. (2003). Factors affecting the early survival and growth of native tree seedlings planted on a degraded hillside grassland in Hong Kong, China. *Restoration Ecology*, 11(4), 483–488.

Hau, B. C. H. and So, K. K. Y. (2003). Using native tree species to restore degraded hillsides in Hong Kong, China. In Sim, H. C., Appanah, S. and Durst, P. B. (Eds.), *The 2002 International Conference on Bringing Back the Forests: Policies and Practices for Degraded Lands and Forests*, Kuala Lumpur, Malaysia, 7–10 October 2002.

Hau, B. C. H. and So, K. K. Y. (2005a). A review of the field performance of native tree-an shrub species planted on man-made slopes in Hong Kong. In *Oral Presentation at Workshop on Concepts and Practices on Slope Bioengineering*, The Chinese University of Hong Kong, 19 November 2005.

Hau, B. C. H. and So, K. K. Y. (2005b). Propagating native tree species for forest rehabitation in Hong Kong, China. In *Proceedings of the Symposium on Tropical Rainforest Rehabilitation and Restoration: Existing Knowledge and Future Direction*, Kota Kinabalu, Malaysia, pp. 26–28.

Hau, B. C. H., So, K. K. Y., Choi, K. C. and Chau, R. Y. H. (2005). Using native tree and shrub species for ecological rehabilitation of man-made slopes in Hong Kong. In *Proceedings of the 25th Annual Seminar*, Geotechnical Division, The Hong Kong Institute of Engineers (Eds.), Hong Kong, China, pp. 273–286.

Heimann, M. and Reichstein, M. (2008). Terrestrial ecosystem carbon dynamics and climate feedbacks. *Nature*, 451(7176), 289–292.

Hemenway, T. (2015). *The Permaculture City: Regenerative Design for Urban, Suburban, and Town Resilience*. Chelsea Green Publishing, White River Junction, VT.

Hillel, D. (1998). *Environmental Soil Physics*. Academic Press, San Diego, CA.

Hinton, E. and Owen, D. R. J. (1979). *An Introduction to Finite Element Computations*. Pineridge Press Ltd., Swansea, UK.

Ho, M. Y. (2007). Governing parameters for stress-dependent soil-water characteristics, conjunctive flow and slope stability. MPhil thesis, The Hong Kong University of Science and Technology, Hong Kong, China.

Hoagland, D. R. and Arnon, D. I. (1950). The water-culture method for growing plants without soil. *California Agricultural Experiment Station Circular*, 347, 1–32.

Höltta, T. and Sperry, J. (2012). Plant water transport and cavitation. In *Proceedings of the NATO Advanced Research Workshop on Alternative Water Resources in Arid Area by Retrieving Water from Secondary Sources*, Ein Bokek, Israel, pp. 173–181.

Hopmans, J. and Bristow, K. (2002). Current capabilities and future needs of root water and nutrient uptake modelling. *Advances in Agronomy*, 77, 104–175.

Hossain, M. A. and Yin, J. H. (2010). Shear strength and dilative characteristics of an unsaturated compacted completely decomposed granite soil. *Canadian Geotechnical Journal*, 47(10), 1112–1126.

Houlsby, G. T. (1991). *How the Dilatancy of Soils Affects Their Behaviour*. University of Oxford, Department of Engineering Science, Oxford, UK.

Hu, Q. M. and Wu, D. L. (2008). *Flora of Hong Kong*, Vol. 2. Agriculture Fisheries and Conservation Department, Hong Kong, China.

Huat, B. B. K., Ali, F. H. J. and Low, T. H. (2006). Water infiltration characteristics of unsaturated soil slope and its effect on suction and stability. *Geotechnical and Geological Engineering*, 24(5), 1293–1306.

Hubble, T., Clarke, S., Stokes, A. and Phillips, C. (2017). Soil bio- and eco-engineering: The use of vegetation to improve slope stability. *Proceedings of the Fourth International Conference. Ecological Engineering*, 109(B), 141–272.

IPCC. (2013). *Climate Change 2013. The Physical Science Basis. Contribution of Working Group I to the Fifth Assessment Report of the Intergovernmental Panel on Climate Change.* Cambridge University Press, Cambridge, UK, p. 1535.

Janott, M., Gayler, S., Gessler, A., Javaux, M., Klier, C. and Priesack, E. (2011). A one-dimensional model of water flow in soil-plant systems based on plant architecture. *Plant Soil,* 341, 233–256.

Jansson, M. (1988). Phosphate uptake and utilization by bacteria and algae. *Hydrobiologia,* 170(1), 177–189.

Jiang, W. S., Wang, K. J., Wu, Q. P., Dong, S. T., Liu, P. and Zhang, J. W. (2013). Effects of narrow plant spacing on root distribution and physiological nitrogen use efficiency in summer maize. *The Crop Journal,* 1(1), 77–83.

Jones, H. G. (2013). *Plants and Microclimate: A Quantitative Approach to Environmental Plant Physiology.* Cambridge University Press, Cambridge, UK.

Jones, H. G. and Rotenberg, E. (2001). *Energy, Radiation and Temperature Regulation in Plants.* In eLS. John Wiley & Sons, Hoboken, NJ.

Jørgensen, B. B. (1977). The sulfur cycle of a coastal marine sediment (Limfjorden Denmark). *Limnology and Oceanography,* 22(5), 814–832.

Joshi, C. P. and Mansfield, S. D. (2007). The cellulose paradox: Simple molecule, complex biosynthesis. *Current Opinion in Plant Biology,* 10, 220–226.

Jotisankasa, A. and Sirirattanachat, T. (2017). Effects of grass roots on soil-water retention curve and permeability function. *Canadian Geotechnical Journal,* 54(11), 1612–1622.

Jotisankasa, A. and Taworn, D. (2016). Direct shear testing of clayey sand reinforced with live stake. *Geotechnical Testing Journal, ASTM,* 39(4), 608–623.

Kamchoom, V., Leung, A. K. and Ng, C. W. W. (2014). Effects of root geometry and transpiration on pull-out resistance. *Geotechnique Letters,* 4(4), 330–336.

Kelliher, F. M., Leuning, R., Raupach, M. R. and Schulze, E. D. (1995). Maximum conduc-tances for evaporation from global vegetation types. *Agricultural and Forest Meteorology,* 73, 1–16.

Kellogg, W. W., Cadle, R. D., Allen, E. R., Lazarus, A. L. and Martell, E. A. (1972). The sulfur cycle. *Science,* 175, 587–596.

Khalili, N. and Khabbaz, M. H. (1998). A unique relationship for χ for the determination of the shear strength of unsaturated soils. *Geotechnique,* 48(2), 1–7.

Klironomos, J. N. (2003). Variation in plant response to native and exotic arbuscular mycorrhizal fungi. *Ecology,* 84, 2292–2301.

Kochummen, K. M., Lafrankie, J. J. V. and Manokaran, N. (1990). Floristic composition of Pasoh Forest Reserve, a lowland rain forest in Peninsula Malaysia. *Journal of Tropical Forest,* 3, 1–13.

Koptur, S., Haber, W. A., Frankie, G. W. and Baker, H. G. (1988). Phenologica studies of shrubs and treelet species in tropical cloud forests of Costa Rica. *Journal of Tropical Ecology,* 4, 323–346.

Köstler, J. N., Bruckner, E. and Bibelriether, H. (1968). *Die Wurzeln der WaldbQume.* Verlag Paul Parey, Hamburg, Germany.

Krahn, J. and Fredlund, D. G. (1972). On total, matric and osmotic suction. *Soil Science,* 114(5), 339–348.

Lal, R. (2004). Soil carbon sequestration impacts on global climate change and food security. *Science,* 304(5677), 1623–1627.

Lam, C. C. and Leung. Y. K. (1995). *Extreme Rainfall Statistics and Design Rainstorm Profiles at Selected Locations in Hong Kong.* Royal Observatory, Hong Kong, China.

Lambers, H., Chapin, F. S. and Pons, T. L. (2008). *Plant Physiological Ecology,* 2nd ed. Springer-Verlag, New York, p. 604, doi:10.1007/978-0-387-78341-3.

Lammertsma, E. I., de Boer, H. J., Dekker, S. C., Dilcher, D. L., Lotter, A. F. and Wagner-Cremer, F. (2011). Global CO_2 rise leads to reduced maximum stomatal conductance in Florida vegetation. *Proceedings of the National Academy of Sciences,* 108(10), 4035–4040.

Langley, J. A. and Megonigal, J. P. (2010). Ecosystem response to elevated CO_2 levels limited by nitrogen-induced plant species shift. *Nature,* 466, 96–99.

Leavitt, S. W. and Danzer, S. R. (1993). Method for batch processing small wood samples to holocel-lulose for stable-carbon isotope analysis. *Analytical Chemistry,* 65, 87–89.

Leung, A. K., Boldrin, D., Liang, T., Wu, Z. Y., Kamchoom, V. and Bengough, A. G. (2017b). Plant age effects on soil infiltration rate during early plant establishment. *Geotechnique*, Ahead of print. doi:10.1680/jgeot.17.T.037.

Leung, A. K., Garg, A., Coo, J. L., Ng, C. W. W. and Hau, B. C. H. (2015b). Effects of the roots of Cynodon dactylon and Schefflera heptaphylla on water infiltration rate and soil hydraulic conductivity. *Hydrological Processes*, 29(15), 3342–3354.

Leung, A. K., Garg, A. and Ng, C. W. W. (2015a). Effects of plant roots on soil-water retention and induced suction in vegetated soil. *Engineering Geology*, 193, 183–197.

Leung, A. K., Kamchoom, V. and Ng, C. W. W. (2017a). Influences of root-induced suction and root geometry on slope stability: A centrifuge study. *Canadian Geotechnical Journal*, 54(3), 291–303.

Leung, A. K. and Ng, C. W. W. (2013a). Analyses of groundwater flow and plant evapotranspiration in a vegetated soil slope. *Canadian Geotechnical Journal*, 50(12), 1204–1218.

Leung, A. K. and Ng, C. W. W. (2013b). Seasonal movement and groundwater flow mechanism in an unsaturated saprolitic hillslope. *Landslides*, 10(4), 455–467.

Leung, A. K. and Ng, C. W. W. (2016). Field investigation of deformation characteristics and stress mobilisation in a soil slope. *Landslides*, 13(2), 229–240.

Leung, A. K., Sun, H. W., Millis, S., Pappin, J. W., Ng, C. W. W. and Wong, H. N. (2011). Field monitoring of an unsaturated saprolitic hillslope. *Canadian Geotechnical Journal*, 48(3), 339–353.

Leung, F. T. Y. (2014). The use of native woody plants in slope upgrading in Hong Kong. PhD thesis, The University of Hong Kong, Hong Kong, China.

Leung, F. T. Y., Yan, W. M., Hau, B. C. H. and Tham, L. G. (2015). Root systems of native shrubs and trees in Hong Kong and their effects on enhancing slope stability. *Catena*, 125, 102–110.

Lewis, J. D., Lucash, M., Olszyk, D. M. and Tingey, D. T. (2002). Stomatal responses of Douglasfir seedlings to elevated carbon dioxide and temperature during the third and fourth years of exposure. *Plant Cell and Environment*, 25, 1411–1421.

Li, Y. and Ghodrati, M. (1994). Preferential transport of nitrate through soil columns containing root channels. *Soil Science Society of American Journal*, 58, 653–659.

Liang, T., Knappett, J. A. and Duckett, N. (2015). Modelling the seismic performance of rooted slopes from individual root–soil interaction to global slope behaviour. *Géotechnique*, 65, 995–1009.

Lilly, S. J. and Currid, P. (2010). *Arborists' Certification Study Guide*. International Society of Arboriculture, Urbana, IL.

Lim, T. T., Rahardjo, H., Chang, M. F. and Fredlund, D. G. (1996). Effect of rainfall on matric suctions in a residual soil slope. *Canadian Geotechnical Journal*, 33(4), 618–628.

Liu, H. W., Feng, S. and Ng, C. W. W. (2016). Analytical analysis of hydraulic effect of vegetation on shallow slope stability with different root architectures. *Computers and Geotechnics*, 80, 115–120.

Loades, K. W., Bengough, A. G., Bransby, M. F. and Hallett, P. D. (2013). Biomechanics of nodal, seminal and lateral roots of barley: Effects of diameter, waterlogging and mechanical impedance. *Plant and Soil*, 370, 407–418.

Loades, K. W., Bengough, A. G., Bransby, M. F. and Hallett, P. D. (2015). Effect of root age on the biomechanics of seminal and nodal roots of barley (Hordeum vulgare L.) in contrasting soil environments. *Plant and Soil*, 395(1–2), 253–261.

Lovelock, J. E., Maggs, R. J. and Rasmussen, R. A. (1972). Atmospheric dimethyl sulphide and the natural sulphur cycle. *Nature*, 237(5356), 452.

Lu, M., Yang, Y., Luo, Y., Fang, C., Zhou, X., Chen, J., Yang, X. and Li, B. (2011). Responses of ecosystem nitrogen cycle to nitrogen addition: A meta-analysis. *New Phytologist*, 189(4), 1040–1050.

Lynch, J. (1995). Root architecture and plant productivity. *Plant Physiology*, 109(1), 7–13.

Ma, R., Cai, C., Li, Z., Wang, J., Xiao, T., Peng, G. and Yang, W. (2015). Evaluation of soil aggregate microstructure and stability under wetting and drying cycles in two Ultisols using synchrotron-based X-ray micro-computed tomography. *Soil and Tillage Research*, 149, 1–11.

Mao, Z., Bourrier, F., Stokes, A. and Fourcaud, T. (2014). Three-dimensional modelling of slope stability in heterogeneous montane forest ecosystems. *Ecological Engineering*, 273, 11–22.

Mao, Z., Saint-André, L., Genet, M., Mine, F. X., Jourdan, C., Rey, H., Courbaud, B. and Stokes, A. (2012). Engineering ecological protection against landslides in diverse mountain forests: Choosing cohesion models. *Ecological Engineering*, 45, 55–69.

Marriott, C. A., Crabtree, J. R., Hood, K. and MacNeil, D. J. (2001). Establishment of vegetation for slope stability. TRL Report 506.

Mattia, C., Bischetti, G. B. and Gentile, F. (2005). Biotechnical characteristics of root systems of typical Mediterranean species. *Plant and Soil*, 278, 23–32.

Maurel, C. and Chrispeels, M. J. (2001). Aquaporins: A molecular entry into plant water relations. *Plant Physiology*, 125(1), 135–138.

Maurel, C., Verdoucq, L., Lu, D.-T. and Santoni, V. (2008). Plant aquaporins: Membrane channels with multiple integrated functions. *Annual Review of Plant Biology*, 59(1), 595–624.

McCarl, B. A., Metting, F. B. and Rice, C. (2006). Soil carbon sequestration. *Climate Change*, 80(1–2), 1–3.

McCully, M. E. (1999). Roots in soil: Unearthing the complexities of roots and their rhizospheres. *Annual Review of Plant Physiology and Plant Molecular Biology*, 50, 695–718.

McElrone, A. J., Choat, B., Gambetta, G. A. and Brodersen, C. R. (2013). Water uptake and transport in vascular plants. *Nature Education Knowledge*, 4(5), 6–18.

Meek, B. D., Rechel, E. R., Carter, L. M., DeTar, W. R. and Urie, A. L. (1992). Infiltration rate of a sandy loam soil: Effects of traffic, tillage, and plant roots. *Soil Science Society of America Journal*, 56(3), 908–913.

Mickovski, S. B., Bengough, A. G., Bransby, M. F., Davies, M. C. R., Hallett, P. D. and Sonnenberg, R. (2007). Material stiffness, branching pattern and soil matric potential affect the pullout resistance of model root systems. *European Journal of Soil Science*, 58(6), 1471–1481.

Mickovski, S. B., Hallett, P. D., Bransby, M. F., Davies, M. C. R., Sonnenberg, R. and Bengough, A. G. (2009). Mechanical reinforcement of soil by willow roots: Impacts of root properties and root failure mechanism. *Soil Science Society of America Journal*, 73, 1276–1285.

Mitchell, A. R., Ellsworth, T. R. and Meek, B. D. (1995). Effect of root systems on preferential flow in swelling soil. *Communications in Soil Science and Plant Analysis*, 26(15–16), 2655–2666.

Moran, L. A., Horton, R. A., Scrimgeour, G., Perry, M. D. and Rawn, J. D. (2011). *Principles of Biochemistry*, 5th ed. Prentice Hall, Boston, MA.

Morris, P. H., Graham, J. and Williams, D. J. (1994). Crack depths in drying clays using fracture mechanics: Fracture mechanics applied to geotechnical engineering. *Geotechnical Special Publication*, 43, 41–53.

Munns, R. and Sharp, R. E. (1993). Involvement of abscisic acid in controlling plant growth in soil of low water potential. *Australian Journal of Plant Physiology*, 20(5), 425–437.

Myers, J. M., Harrison, D. S. and Phillips, J. W. (1976). Soil moisture distribution in a sprinkler irrigated orange grove. *Proceedings of the Florida State Horticultural Society*, 89, 23–28.

Ng, C. W. W. (2007). Keynote paper: Liquefied flow and non-liquefied slide of loose fill slopes. In *Proceedings of the 13th Asian Regional Conference on Soil Mechanics and Geotechnical Engineering*, 10–14 December, Allied Publishers Private Ltd, Kolkata, India. Vol. 2 (post-conference volume), pp. 120–134.

Ng, C. W. W. (2009). What is static liquefaction failure of loose fill slope. In *The 1st Italian Workshop on Landslides*, Napoli, Italy, 1, 91–102.

Ng, C. W. W. (2014). The 6th ZENG Guo-xi Lecture: The state-of-the-art centrifuge modelling of geotechnical problems at HKUST. *Journal of Zhejiang University-Science A (Applied Physics & Engineering)*, 15(1), 1–21.

Ng, C. W. W. (2017). Atmosphere-plant-soil interaction: Theories and mechanisms. *Chinese Journal of Geotechnical Engineering*, 39(1), 1–47.

Ng, C. W. W., Coo, J. L., Chen, Z. K. and Chen, R. (2016b). Water infiltration into a new three-layer landfill cover system. *Journal of Environmental Engineering ASCE*, 142(5), 04016007.

Ng, C. W. W., Garg, A., Leung, A. K. and Hau, B. C. H. (2016c). Relationships between leaf and root area indices and soil suction induced during drying-wetting cycles. *Ecological Engineering*, 91, 113–118.

Ng, C. W. W., Kamchoom, V. and Leung, A. K. (2016a). Centrifuge modelling of the effects of root geometry on the transpiration-induced suction and stability of vegetated slopes. *Landslides*, 13(5), 1–14.

Ng, C. W. W. and Leung, A. K. (2012). Measurements of drying and wetting permeability functions using a new stress-controllable soil column. *Journal of Geotechnical and Geoenvironmental Engineering, ASCE*, 138(1), 58–68.

Ng, C. W. W., Leung, A. K., Kamchoom, V. and Garg, A. (2014b). A novel root system for simulating transpiration-induced soil suction in centrifuge. *Geotechnical Testing Journal*, 37(5), 1–15.

Ng, C. W. W., Leung, A. K. and Ni, J. J. (2018a). Bioengineering for slope stabilisation using plants: Hydrological and mechanical effects. In *Proceedings of China-Europe Conference on Geotechnical Engineering*, Springer, Cham, Switzerland, pp. 1287–1303.

Ng, C. W. W., Leung, A. K. and Woon, K. X. (2014a). Effects of soil density on grass-induced suction distributions in compacted soil subjected to rainfall. *Canadian Geotechnical Journal*, 51(3), 311–321.

Ng, C. W. W., Liu, H. W. and Feng, S. (2015). Analytical solutions for calculating pore water pressure in an infinite unsaturated slope with different root architectures. *Canadian Geotechnical Journal*, 52(12), 1981–1992.

Ng, C. W. W. and Menzies, B. (2007). *Advanced Unsaturated Soil Mechanics and Engineering*. Taylor & Francis Group, London, UK.

Ng, C. W. W., Ni, J. J., Leung, A. K. and Wang, Z. J. (2016d). A new and simple water retention model for root-permeated soils. *Géotechnique Letters*, 6(1), 106–111.

Ng, C. W. W., Ni, J. J., Leung, A. K., Zhou, C. and Wang, Z. J. (2016e). Effects of planting density on tree growth and induced soil suction. *Géotechnique*, 66(9), 711–724.

Ng, C. W. W. and Pang, Y. W. (2000a). Experimental investigations of the soil-water characteristics of a volcanic soil. *Canadian Geotechnical Journal*, 37(6), 1252–1264.

Ng, C. W. W. and Pang, Y. W. (2000b). Influence of stress state on soil-water characteristics and slope stability. *Journal of Geotechnical and Geoenvironmental Engineering, ASCE*, 126(2). doi:10.1061/(ASCE)1090-0241(2000)126:2(157).

Ng, C. W. W., Pun, W. K., Kwok, S. S. K., Cheuk, C. Y. and Lee, D. M. (2007). Centrifuge modelling in engineering practice in Hong Kong. In *Proceeding of the Geotechnical Division's Annual Seminar*, Hong Kong Institution of Engineers (HKIE), pp. 55–68.

Ng, C. W. W., Tasnim, R. and Wong, J. J. F. (2018b). Coupled effects of atmospheric CO_2 concentration and nutrients on plant-induced soil suction. Submitted to Geotechnique Letters.

Ng, C. W. W., Van Laak, P., Tang, W. H., Li, X. S. and Zhang, L. M. (2001). The Hong Kong geotechnical centrifuge. In *Proceedings of 3rd International Conference on Soft Soil Engineering*, Hong Kong, China, pp. 225–230.

Ng, C. W. W., Van Laak, P. A., Zhang, L. M., Tang, W. H., Zong, G. H., Wang, L. Z., Xu, G. M. et al. (2002). Development of a four-axis robotic manipulator for centrifuge modelling at HKUST. In *Proceedings of International Conference on Physical Modelling in Geotechnics*, St. John's Newfoundland, Canada, pp. 71–76.

Ng, C. W. W., Wang, Z. J., Leung, A. K. and Ni, J. J. (2018c). A study on effects of leaf and root characteristics on plant root water uptake. *Géotechnique*. doi:10.1680/jgeot.16.p.332.

Ng, C. W. W., Wong, H. N., Tse, Y. M., Pappin, J. W., Sun, H. W., Millis, S. W. and Leung, A. K. (2011). A field study of stress-dependent soil-water characteristic curves and permeability of a saprolitic slope in Hong Kong. *Geotechnique*, 61(6), 511–521.

Ng, C. W. W., Woon, K. X., Leung, A. K. and Chu, L. M. (2013). Experimental investigation of induced suction distributions in a grass-covered soil. *Ecological Engineering*, 52, 219–223.

Ng, C. W. W. and Zhan, L. T. (2007). Comparative study of rainfall infiltration into a bare and a grassed unsaturated expansive soil slope. *Soils and Foundations*, 47(2), 207–217.

Ng, C. W. W., Zhan, L. T., Bao, C. G., Fredlund, D. G. and Gong, B. W. (2003). Performance of an unsaturated expansive soil slope subjected to artificial rainfall infiltration. *Géotechnique*, 53(2), 143–157.

Ng, C. W. W. and Zhou, R. Z. B. (2005). Effects of soil suction on dilatancy of an unsaturated soil. In *Proceedings of the 16th International Conference on Soil Mechanics and Geotechnical Engineering*. Osaka, Japan, pp. 559–562.

Ng'etich, O., Niyokuri, O. K. A. N., Rono, J. J., Fashaho, A. and Ogweno, J. O. (2013). Effect of different rates of nitrogen fertilizer on the growth and yield of zucchini (Cucurbita pepo cv. Diamant L.) Hybrid F1 in Rwandan high-altitude zone. *International Journal of Agriculture and Crop Sciences*, 5(1), 54–62.

Ni, J. J., Leung, A. K. and Ng, C. W. W. (2017). Investigation of plant growth and transpiration-induced suction under mixed grass-tree conditions. *Canadian Geotechnical Journal*, 54(4), 561–573.

Ni, J. J., Leung, A. K. and Ng, C. W. W. (2018a). Modelling soil suction changes due to mixed species planting. *Ecological Engineering*, 117, 1–17.

Ni, J. J., Leung, A. K., Ng, C. W. W. and Shao, W. (2018b). Modelling hydro-mechanical reinforcements of plants to slope stability. *Computers and Geotechnics*, 95, 99–109.

Niklas, K. J. (1992). *Plant Biomechanics: An Engineering Approach to Plant Form and Function.* University of Chicago Press, Chicago, IL.

Nilaweera, N. S. and Nutalaya, P. (1999). Role of tree roots in slope stabilisation. *Bulletin of Engineering Geology and the Environment*, 57(4), 337–342.

Nobel, P. S. (2009). *Physicochemical & Environmental Plant Physiology.* Academic Press, London, UK.

Norris, J. E., Di Iorio, A., Stokes, A., Nicoll, B. C. and Achim, A. (2008). Species selection for soil reinforcement and protection. In *Slope Stability and Erosion Control: Ecotechnological Solutions*, Norris, J., Stokes, A., Mickovski, S., Cammeraat, E., van Beek, R., Nicoll, B. and Achim, A. (Eds.). Springer, Dordrecht, the Netherlands, pp. 167–210.

Or, I. O. L., Hau, B. C. H. and Cheung, R. W. M. (2011). Application of native plant species in the landslip preventive measures programme. In *Proceedings of the 31st Annual Seminar*, Geotechnical Division, Hong Kong Institute of Engineers, Hong Kong, China, pp. 163–169.

Osman, N. and Barakbah, S. (2011). The effect of plant succession on slope stability. *Ecological Engineering*, 37, 139–147.

Osman, N. and Barakbah, S. S. (2006). Parameters to predict slope stability-soil water and root profiles. *Ecological Engineering*, 28, 90–95.

Partov, D., Maślak, M., Ivanov, R., Petkov, M., Sergeev, D. and Dimitrova, A. (2016). The development of wooden bridges through the ages: A review of selected examples of heritage objects, Part I – The milestones. *Technical Transactions*, 2-B, 93–106.

Pauly, M., Gille, S., Liu, L., Mansoori, N., Souza, A., de Schultink, A. and Xiong, G. (2013). Hemicellulose biosynthesis. *Planta*, 238, 627–642.

Peng, X., Horn, R. and Smucker, A. (2007). Pore shrinkage dependency of inorganic and organic soils on wetting and drying cycles. *Soil Science Society of America Journal*, 71(4), 1095–1103.

Penman, H. L. (1948). Natural evaporation from open water, bare soil and grass. In *Proceedings of the Royal Society of London A*, 193(1032), 120–145.

Pérez-Harguindeguy, N., Díaz, S., Garnier, E., Lavorel, S., Poorter, H. et al. (2013). New handbook for standardised measurement of plant functional traits worldwide. *Australian Journal of Botany*, 61, 167–234.

Pollen, N. and Simon, A. (2005). Estimating the mechanical effects of riparian vegetation on stream bank stability using a fiber bundle model. *Water Resources Research*, 41, W07025.

Polyanin, A. D. (2002). *Linear Partial Differential Equations for Engineers and Scientists*, Chapman and Hall/CRC Press, Boca Raton, FL.

Poorter, H., Niinemets, U., Poorter, L., Wright, I. J. and Villar, R. (2009). Causes and consequences of variation in leaf mass per area (LMA): A meta-analysis. *New Phytologist*, 182, 565–588.

Postgate, J. (1998). *Nitrogen Fixation*, 3rd ed. Cambridge University Press, Cambridge, UK.

Prasad, R. (1988). A linear root water uptake mode. *Journal of Hydrology*, 99(3), 297–306.

Preti, F. (2013). Forest protection and protection forest: Tree root degradation over hydrological shallow landslides triggering. *Ecological Engineering*, 61, 633–645.

Preti, F. and Giadrossich, F. (2009). Root reinforcement and slope bioengineering stabilization by Spanish Broom (*Spartium junceum* L.). *Hydrology and Earth System Sciences*, 13(9), 1713–1726.

Preti, F. and Schwarz, M. (2006). On root reinforcement modelling. Geophysical Research 436 Abstracts, vol. 8, EGU General Assembly 2006, 2–7 April, ISSN: 1029-7006.

Qiu, Z. C., Wang, M., Lai, W. L., He, F. H. and Chen, Z. H. (2011). Plant growth and nutrient removal in constructed monoculture and mixed wetlands related to stubble attributes. *Hydrobiologia*, 661(1), 251–260.

Raats, P. A. C. (1974). Steady flows of water and salt in uniform soil profiles with plant roots. *Soil Science Society of America Journal*, 38(5), 717–722.

Rahardjo, H., Satyanaga, A., Leong, E. C., Santoso, V. A. and Ng, Y. S. (2014). Performance of an instrumented slope covered with shrubs and deep-rooted grass. *Soils and Foundations*, 54(3), 417–425.

Reich, P. B., Hobbie, S. E., Lee, T., Ellsworth, D. S., West, J. B., Tilman, D., Knops, J. M., Naeem, S. and Trost, J. (2006). Nitrogen limitation constrains sustainability of ecosystem response to CO_2. *Nature*, 440, 922–925.

Reich, P. B., Walters, M. B. and Ellsworth, D. S. (1997). From tropics to tundra: Global convergence in plant functioning. *Proceedings of the National Academy of Sciences of the United States of America*, 94, 13730–13734.

Richards, L. (1965). Measuring of the free energy of soil moisture by psychrometric technique using thermistors. *Psychrometers*, 39–46.

Rieger, M. and Litvin, P. (1999). Root system hydraulic conductivity in species with contrasting root anatomy. *Journal of Experimental Botany*, 50(331), 201–209.

Roberts, J., Jackson, N. and Smith, M. (2006). *Tree Roots in the Built Environment (No. 8)*. The Stationery Office, London, UK.

Romero, E., Gens, A. and Lloret, A. (1999). Water permeability, water retention and microstructure of unsaturated compacted boom clay. *Engineering Geology*, 54(1–2), 117–127.

Roose, T. and Fowler, A. C. (2004). A model for water uptake by plant roots. *Journal of Theoretical Biology*, 228, 155–171.

Rozados-Lorenzo, M. J., González-Hernádez, M. P. and Silva-Pando, F. J. (2007). Pasture production under different tree species and densities in an Atlantic silvopastoral system. *Agroforestry Systems*, 70(1), 53–62.

Sack, L. and Tyree, M. T. (2005). *Leaf Hydraulics and Its Implications in Plant Structure and Function: Vascular Transport in Plants*. Elsevier Academic Press, London, UK, pp. 93–114.

Saifuddin, M., Osman, N., Motior Rahman, M. and Boyce, A. N. (2015). Soil reinforcement capability of two legume species from plant morphological traits and mechanical properties. *Current Science*, 108(7), 1340–1347.

Saxena, I. M. and Brown, R. M. (2005). Cellulose biosynthesis: Current views and evolving concepts. *Annals of Botany*, 96, 9–21.

Scanlan, C. A. and Hinz, C. (2010). Insight into the processes and effects of root induced changes to soil hydraulic properties. In *19th World Congress of Soil Science, Soil Solutions for a Changing World*, Brisbane, Australia, 1–6 August 2010, pp. 41–44.

Schachtman, R., Reid, R. J. and Ayling, S. M. (1998). Phosphorus uptake by plants: From soil to cell. *Plant Physiology*, 116(2), 447–453.

Scheller, H. V. and Ulvskov, P. (2010). Hemicelluloses. *Annual Review of Plant Biology*, 61, 263–289.

Schimel, D. S. (1995). Terrestrial ecosystems and the carbon cycle. *Global Change Biology*, 1(1), 77–91.

Scholes, R. J. and Archer, S. R. (1997). Tree-grass interactions in savannas. *Annual Review of Ecology and Systematics*, 28, 517–544.

Scholl, P., Leitner, D., Kammerer, G., Lioskandl, W., Kaul, H. P. and Bodner, G. (2014). Root induced changes of effective 1D hydraulic properties in a soil column. *Plant and Soil*, 381(1–2), 193–213.

Shao, W., Ni, J. J., Leung, A. K., Su, Y. and Ng, C. W. W. (2017). Analysis of plant root-induced preferential flow and pore water variation by a dual-permeability model. *Canadian Geotechnical Journal*, 54(11), 1537–1552.

Shen, C. K., Li, X. S., Ng, C. W. W., Van Laak, P. A., Kutter, B. L., Cappel, K. and Tauscher, R. C. (1998). Development of a geotechnical centrifuge in Hong Kong. In *Proceedings of Centrifuge*, Tokyo, Japan, pp. 13–18.

Siau, J. F. (1984). *Transport Processes in Wood*. Springer, Berlin, Germany.

Simon, A. and Collison, A. J. C. (2002). Quantifying the mechanical and hydrological effects of riparian vegetation on streambank stability. *Earth Surface Processese and Landforms*, 27(5), 527–546.

Skerman, P. J. and Riveros, F. (1990). *Tropical Grasses*. FAO of the United Nations, Rome, Italy.

Smith, I. and Snow, M. (2008). Timber: An ancient construction material with a bright future. *Forestry Chronicle*, 4(84), 504–510.

Smith, S. E. and Read, D. J. (2008). *Mycorrhizal Symbiosis*, 3rd ed. Academic Press, London, UK.

Sonnenberg, R., Bransby, M. F., Bengough, A. G., Hallett, P. D. and Davies, M. C. R. (2012). Centrifuge modelling of soil slopes containing model plant roots. *Canadian Geotechnical Journal*, 49(1), 1–17.

Sonnenberg, R., Bransby, M. F., Hallett, P. D., Bengough, A. G., Mickovski, S. B. and Davies, M. C. R. (2010). Centrifuge modelling of soil slopes reinforced with vegetation. *Canadian Geotechnical Journal*, 47(12), 1415–1430.

Sperry, J. S., Adler, F. R., Campbell, G. S. and Comstock, J. P. (1998). Limitation of plant water use by rhizosphere and xylem conductance: Results from a model. *Plant, Cell & Environment*, 21, 347–359.

Stepniewski, W., Glinski, J. and Ball, B. C. (1994). Effects of compaction on soil aeration properties. In *Soil Compaction in Crop Production*, Soane, B. D. and van Ouwerkerk, C. (Eds.), Elsevier, Amsterdam, the Netherlands, pp. 167–190.

Steudle, E. (2000). Water uptake by plant roots: An integration of views. *Plant and Soil*, 226(1), 45–56.

Steudle, E. (2001). The cohesion-tension mechanism and the acquisition of water by plant roots. *Annual Review of Plant Biology*, 52, 847–875.

Stevenson, F. J. and Cole, M. A. (2008). *Cycles of Soils: Carbon, Nitrogen, Phosphorus, Sulfur, Micronutrients*, 2nd ed. Wiley, New York.

Stokes, A., Atger, C., Bengough, A. G., Fourcaud, T. and Sidle, R. C. (2009). Desirable plant root traits for protecting natural and engineered slopes against landslides. *Plant and Soil*, 324, 1–30.

Stokes, A. and Mattheck, C. (1996). Variation of wood strength in tree roots. *Journal of Experimental Botany*, 47(5), 693–699.

Stokes, A., Norris, J. E., van Beek, L. P. H., Bogaard, T., Cammeraat, E., Mickovski, S. B., Jenner, A. et al. (2008). How vegetation reinforces soil on slopes. In: *Slope Stability and Erosion Control: Ecotechnological Solutions*, pp. 65–118.

Stokes, A., Spanos, I., Norris, J. E. and Cammeraat, E. (Eds.). (2007). Eco-and ground bio-engineering: The use of vegetation to improve slope stability. In *Proceedings of the First International Conference on Eco-engineering*, 13–17 September 2004, Springer.

Stokes, I. S., Norris, J. E. and Cammeraat, E. (2004). *Eco-and Ground Bio-Engineering: The Use of Vegetation to Improve Slope Stability*. Springer, Dordrecht, the Netherlands, pp. 213–221.

Świtała, B. M., Askarinejad, A., Wu, W. and Springman, S. M. (2018). Experimental validation of a coupled hydro-mechanical model for vegetated soil. *Géotechnique*, 68(5), 375–385.

Taylor, N. G. (2008). Cellulose biosynthesis and deposition in higher plants. *New Phytologist*, 178, 239–252.

Taylor, R. N. (1995). *Geotechnical Centrifuge Technology*. Taylor & Francis Group, London, UK.

Thetaprobe. (1999). *Theta-probe: Soil Moisture Sensor*. Delta-T Device Ltd, Cambridge, UK.

Thrower, S. L. (1984). *Hong Kong Shrub*, Vol. II. The Urban Council of Hong Kong, Hong Kong, China.

Tomlinson, M. and Woodward, J. (2014). *Pile Design and Construction Practice*, 6th ed. CRC Press, Taylor & Francis Group, Boca Raton, FL.

Tosi, M. (2007). Root tensile strength relationships and their slope stability implications of three shrub species in the Northern Apennines (Italy). *Geomorphology*, 87, 268–283.

Tyree, M. T. and Ewers, F. W. (1991). The hydraulic architecture of trees and other woody plants. *New Phytologist*, 119, 345–360.

USGS. (2016). The water cycle. U. S. Geological Survey, USA. http://water.usgs.gov/edu/watercycle.html.

van Genuchten, M. T. (1980). A closed-form equation for predicting the hydraulic conductivity of unsaturated soils. *Soil Science Society of America Journal*, 44(5), 892–898.

van Noordwijk, M., Widianto, H. M. and Hairah, K. (1991). Old tree root channels in acid soils in the humid tropics: Important for crop root penetration, water infiltration and nitrogen management. *Developments in Plant and Soil Sciences*, 134(1), 37–44.

Vanapalli, S. K., Fredlund, D. G., Pufahi, D. E. and Clifton, A. W. (1996). Model for the prediction of shear strength with respect to soil suction. *Canadian Geotechnical Journal*, 33, 379–392.

Veiga, R. S. L., Faccio, A., Genre, A., Pieterse, C. M. J., Bonfante, P. and van der Heijden, M. G. A. (2013). Arbuscular mycorrhizal fungi reduce growth and infect roots of the non-host plant Arabidopsis thaliana. *Plant, Cell & Environment*, 36, 1926–1937.

Vergani, C. and Graf, F. (2015). Soil permeability, aggregate stability and root growth: A pot experiment from a soil bioengineering perspective. *Ecohydrology*. doi:10.1002/eco.1686.

Vergani, C., Schwarz, M., Cohen, D., Thormann, J. J. and Bischetti, G. B. (2014). Effects of root tensile force and diameter distribution variability on root reinforcement in the Swiss and Italian alps. *Canadian Journal of Forest Research*, 44(11), 1426–1440.

Waldron, L. J. and Dakessian, S. (1981). Soil reinforcement by roots: Calculation of increased soil shear resistance from root properties. *Soil Science*, 132(6), 427–435.

Washbourne, C. L., Lopez-Capel, E., Renforth, P., Ascough, P. L. and Manning, D. A. C. (2015). Rapid removal of atmospheric CO_2 by urban soils. *Environmental Science & Technology*, 49(9), 5434–5440.

Watson, A., Phillips, C. and Marden, M. (1999). Root strength, growth, and rates of decay: Root reinforcement changes of two tree species and their contribution to slope stability. *Plant and Soil*, 217(1–2), 39–47.

Watson, D. J. (1947). Comparative physiological studies on the growth of field crops: I. Variation in net assimilation rate and leaf area between species and varieties and within and between years. *Annals of Botany*, 11(1), 41–76.

Watson, G. W. (1987). The relationship of root growth and tree vigour following transplanting. *Arboricultural Association Journal*, 11(2), 97–104.;

Watson, K. K. (1966). An instantaneous profile method for determining the hydraulic conductivity of unsaturated porous materials. *Water Resources Research*, 2(4): 709–715.

Wheeler, T. D. and Stroock, A. D. (2008). The transpiration of water at negative pressures in a synthetic tree. *Nature*, 455(7210), 208–212.

White, D. J., Take, W. A. and Bolton, M. D. (2003). Soil deformation measurement using particle image velocimetry (PIV) and photogrammetry. *Geotechnique*, 53(7), 619–631.

White, P. J. and Brown, P. H. (2010). Plant nutrition for sustainable development and global health. *Annals of Botany*, 339(1–2), 1–2.

Wightman, R. and Turner, S. (2010). Trafficking of the plant cellulose synthase complex. *Plant Physiology*, 153, 427–432.

Wong, C. C., Wu, S. C., Kuek, C., Khan, A. G. and Wong, M. H. (2007). The role of mycorrhizae associated with vetiver grown in Pb-/Zn-contaminated soils: Greenhouse study. *Restoration Ecology*, 15, 60–67.

Woon, E. (2013). Field and laboratory investigations of Bermuda grass induced suction and distribution. MPhil thesis, The Hong Kong University of Science and Technology, Hong Kong, China.

Wright, I. J. O., Reich, P. B. O., Westoby, M. O., Ackerly, D. D., Baruch, Z., Bongers, F., Cavender-Bares, J. O. et al. (2004). The worldwide leaf economics spectrum. *Nature*, 428, 821–827.

Wu, Q. S., Liu, C. Y., Zhang, D. J., Zou, Y. N., He, X. H. and Wu, Q. H. (2016). Mycorrhiza alters the profile of root hairs in trifoliate orange. *Mycorrhiza*, 26(3), 237–247.

Wu, T. H., McKinnell, W. P. and Swanston, D. N. (1979). Strength of tree roots and landslides on Prince of Wales Island, Alaska. *Canadian Geotechnical Journal*, 16(1), 19–33.

Yan, W. M. and Zhang, G. H. (2015). Soil-water characteristics of compacted sandy and cemented soils with and without vegetation. *Canadian Geotechnical Journal*, 52(9), 1–14.

Yang, Y., Chen, L., Li, N. and Zhang, Q. (2016). Effect of root moisture content anddiameter on root tensile properties. *PLoS One*, 11(3), e0151791.

Yin, J. H. (2009). Influence of relative compaction on the hydraulic conductivity of completely decomposed granite in Hong Kong. *Canadian Geotechnical Journal*, 46(10), 1229–1235.

Yuan, F. and Lu, Z. (2005). Analytical solutions for vertical flow in unsaturated, rooted soils with variable surface fluxes, *Vadose Zone Journal*, 4(4), 1210–1218.

Zhai, Y., Law, K. T. and Lee, C. F. (2000). Shear behaviour of CDG loose fill under undrained triaxial compression. In *Symposium on Slope Hazards and Their Prevention*, pp. 338–343.

Zhan, L. T. (2003). Field and laboratory study of an unsaturated expansive soil associated with rain-induced slope instability. PhD thesis, The Hong Kong University of Science and Technology.

Zhan, T. L. T., Jia, G. W., Chen, Y. M., Fredlund, D. G. and Li, H. (2013). An analytical solution for rainfall infiltration into an unsaturated infinite slope and its application to slope stability analysis. *International Journal for Numerical and Analytical Methods in Geomechanics*, 37(12), 1737–1760.

Zhang, C. B., Chen, L. H. and Jiang, J. (2014). Why fine tree roots are stronger than thicker roots: The role of cellulose and lignin in relation to slope stability. *Geomorphology*, 206, 196–202.

Zhang, M. (2006). Centrifuge modelling of potentially liquefiable loose fill slopes with and without soil nails. PhD thesis, The Hong Kong University of Science and Technology, Hong Kong, China.

Zhang, M. and Ng, C. W. W. (2003). *Interim Factual Testing Report I-SG30 & SR30*. The Hong Kong University of Science and Technology, Hong Kong, China.

Zhang, M., Ng, C. W. W., Take, W. A., Pun, W. K., Shiu, Y. K. and Chang, G. W. K. (2006). The role and mechanism of soil nails in liquefied loose sand fill slopes. In *Proceedings of 6th International Conference Physical Modelling in Geotechnics*, Hong Kong, China, pp. 391–396.

Zhou, Z. B. (2008). Centrifuge and three-dimensional numerical modelling of steep CDG slopes reinforced with different sizes of nail heads. PhD thesis, The Hong Kong University of Science and Technology.

Author index

Subject index

Printed in the United States
by Baker & Taylor Publisher Services